Specifications for Building Conservation

In *Specifications for Building Conservation*, the National Trust draws on a range of case studies and specifications to provide a much needed guide to specification writing for building conservation. Although traditional building accounts for approximately a quarter of all buildings in the UK, the old skills and understanding required for their care and maintenance have been increasingly eroded over the last century. As the largest heritage charity in Europe, the National Trust has a first-class reputation for high standards of conservation and care, and in this three-volume set, the Trust brings together a remarkable pool of expertise to guide conservation professionals and students through the process of successful specification writing.

This first book focuses on the materials used for the external fabric, detailing successful approaches employed by the National Trust at some of its most culturally significant sites. A range of studies have been carefully selected for their interest, diversity and practicality, showcasing projects from stonework repairs on the magnificent Grade I-listed Hardwick Hall to the rethatching of the traditional cottages of the Holnicote Estate.

Complete with a practical conservation management plan checklist, this book will enable practitioners to develop their skills, allowing them to make informed decisions when working on a range of project types. This is the first practical guide to specification writing for building conservation and the advice provided by the National Trust experts will be of interest to any practitioners and students involved in building conservation, both in the UK and beyond.

The National Trust was founded in 1895 with the aim of caring for special places, for ever, for everyone. Now, over 120 years later, we look after historic houses and gardens, industrial monuments and mills, archaeological sites, nature reserves, coastlines, forests, and more. We have a portfolio of over 28,000 buildings and structures, across England, Wales and Northern Ireland. They include over 300 mansions, 57 entire villages, over 5000 cottages (some tenanted, some used as holiday cottages), farms, bridges, lighthouses and many others. We have built a formidable reputation for the conservation and care of places of natural beauty and historic significance. Even more importantly, we make sure that our places are available to all, with over 17 million visitors to pay-for-entry properties each year. We have 4.4 million members and 62,000 volunteers, who are vital to our work. We are a registered charity and are completely independent of Government, relying on membership fees, donations and legacies, and revenue raised from our commercial operations.

'This volume fills a gap in the current literature by providing an immensely practical guide to specification writing for building conservation. Drawing on a broad range of National Trust projects, it will satisfy conservation professionals, students, and all those with a thirst for technical detail by sharing the tried and tested specifications but more importantly explaining the underlying conservation philosophy in each case.'
Kate Gunthorpe, MRICS BCAS, Senior Building Surveyor, Historic England

Specifications for Building Conservation

Volume 1: External structure

Edited by Rory Cullen and Rick Meier

Routledge
Taylor & Francis Group

LONDON AND NEW YORK

National
Trust

First published 2016 by Routledge

2 Park Square, Milton Park, Abingdon, Oxon OX14 4RN

605 Third Avenue, New York, NY 10017

First issued in paperback 2021
First issued in hardback 2019

Routledge is an imprint of the Taylor & Francis Group, an informa business

British Library Cataloguing-in-Publication Data
A catalogue record for this book is available from the British Library

Library of Congress Cataloging in Publication Data
Specifications for building conservation.
volumes cm
Contents: Volume 1. External structure --
Includes bibliographical references.
1. Historic buildings--Conservation and restoration--Specifications--Great Britain.
2. Buildings--Repair and reconstruction--Specifications. 3. Specification writing.
I. National Trust (Great Britain)
TH425.S653 2015
692'.3--dc23
2015007556

ISBN 13: 978-1-03-209835-7 (pbk)
ISBN 13: 978-1-873394-80-9 (hbk)

Typeset in Syntax by Servis Filmsetting Ltd, Stockport, Cheshire

Contents

Preface

Specifiers typically have to rely on their own experience in the preparation of a specification, often adapting earlier similar documents from previous projects. The consequence can be a disjointed and/or ambiguous specification, with inappropriate materials and techniques, which the contractor is left to translate into practice using their own experience and understanding. In too many cases, neither specifier nor contractor may have the knowledge or understanding of traditional building skills and techniques to make appropriate judgements.

Modern new-build technology is used for the vast majority of building works in the UK and training in writing specifications focuses on this, ignoring the fact that traditional building stock accounts for approximately a quarter of all buildings. The care and repair of traditional buildings require a very different approach to modern constructions, because the structure and materials perform in a different manner. The old skills and understanding required have been increasingly eroded over the last century, evidenced in the poor repair and maintenance of many historic buildings.

The National Trust was founded in 1895 with the aim of conserving places of historic interest or natural beauty. It has a formidable reputation for high standards of conservation and care. Today, it looks after an estimated 28,000 historic buildings and structures spanning many centuries of construction technique. As the largest heritage charity in Europe, the Trust is in a prime position to share its knowledge and experience in the building conservation field. It supports and develops the retention and growth of traditional skills through apprenticeship schemes and training, as well as seeking to share good practice within the heritage and building sectors.

This, therefore, is the first of three volumes in a series on specification writing for conservation. This volume focuses on the materials used in the construction of the external elements of a building; the next two volumes will cover materials, maintenance and cleaning, plus internal fabric. All three books will follow the same framework – an explanation of the project, the conservation philosophy considered and the specification or schedule of works; technology is largely excluded as it advances so rapidly and may quickly lose relevance. The books are not intended to be a definitive guide of how to write a specification for historic buildings, but aim to showcase some of the successful approaches that have been employed by the Trust.

By using case studies and actual specifications, mostly of places that can be visited, it is possible to see first-hand how the Trust approaches its projects and gain a better insight into how ideas and methods could be transferred to other building or conservation projects. Although historic building specifications can be found in the libraries of conservation organisations and offices of surveyors, architects and engineers, no-one has produced such a series of case studies for industry-wide use. Whilst these professions are likely to be the primary audience, these books will be of equal interest to others, including students and owners or custodians of historic buildings.

These specifications have been compiled with the intention that readers will find their contents interesting, stimulating and, above all, useful. They have been written by many different staff and consultants, and we hope that their endeavours will be rewarded by the interest in their work. The examples selected form a cross-section of projects demonstrating the various ways in which a specification can be constructed; well-annotated sketches or diagrams, for instance, can be every bit as valuable as a highly detailed written technical specification. This range of approaches helps to ensure that contractors are fully aware of how they need to undertake the work and can price accurately and on a comparative basis.

We would like to thank the many staff and consultants who have contributed their specifications for use in this publication. We would also like to thank Jill Pearce for her faith, patience and forbearance in the long road to the publication of these books, and Mike Fitzgerald, who was heavily involved in the initial stages of research.

Above all, however, we owe a huge debt of gratitude to the voluntary efforts of Rick Meier, who cajoled all those involved into providing their specifications, and then threw himself into the long-winded editorial task which has culminated in this series, all of which was undertaken with his customary good humour.

Rory Cullen
Head of Buildings
National Trust

Introduction

A good specification should provide a clear methodology for a contractor to understand the requirements of the design. This will ensure that the work can be accurately priced and undertaken correctly. It should be organised in a logical manner, leaving no room for confusion or misinterpretation.

Part 1 provides a concise overview of the specification process within historic buildings:

- importance of conservation philosophy
- heritage legislation
- procurement processes
- types of specifications that might be considered
- the different components these involve.

Part 2 comprises project case studies. It begins with research into building recording and then introduces different building materials.

- *Building recording* – assessing the structure to determine its significance in order to inform building conservation specifications.
- *Masonry: brickwork and stonework* – from the grandeur of Hardwick Hall to a small Cornish vernacular structure and the intricate detailing of the brick chimney stacks at Thorington Hall.
- *Timber repairs* – complex repairs to the gatehouse for Lower Brockhampton House and to a donkey wheel on a Sussex farm.
- *Stone roofing coverings* – from post-fire repairs at the Fleece Inn, Bretforton, to unusual fish-scale slating at Belton House's boathouse and Hamstone slates at Lytes Cary Manor.
- *Thatch* – focusing on long-straw thatch on a 16th-century Hertfordshire cottage and combed wheat reed on the Holnicote Estate.

Each section begins with an introduction outlining the subject matter, the particular projects and the conservation philosophy employed, before describing the specification itself with comments on the results. We have included images, drawings and plans where available to provide a context for each project.

Biographies

RICK MEIER

Rick Meier spent two previous lives in the software industry before joining the historic building conservation community. Rick is a volunteer for the National Trust working primarily with the Small Buildings Survey Team in the South East Region. He has also worked on special projects for the Trust's Buildings Department. Rick has a BSc in Building Surveying and is a member of The Society for the Protection of Ancient Buildings, The Vernacular Architecture Group and The Domestic Buildings Research Group (Surrey).

RORY CULLEN

Rory Cullen has been Head of Buildings at the National Trust since October 2002. A key part of this role is to ensure consistent best practice approaches across the National Trust on all aspects of building work. Externally he is chairman of the Chartered Institute of Building Conservation MARC Group (Maintenance, Adaptation, Refurbishment and Adaptation), sits on the Editorial Board for the Journal of Architectural Conservation, is a Trustee of COTAC, and is a qualified assessor for high-level Conservation NVQs. He is an assessor for high-level NVQ qualifications, and other qualifications include an MSc in Building Conservation, membership of the Institute of Historic Building Conservation and Fellowship of the Chartered Institute of Building.

NIGEL HOUGHTON

Nigel Houghton has been a Building Surveyor at the National Trust since August 2003. His role is that of providing professional building surveying and project management services in connection management and delivery of conservation and development building works. His main area of interest (and also a common challenge of the work for the National Trust) is exploring and working with the challenge between the commercial and environmental sustainability requirements, and the need to preserve and conserve buildings/sites to provide a sustainable basis for their use. His qualifications include an MSc in Sustainable Heritage, an NVQ in Conservation Consultancy, and membership of the Chartered Institute of Building.

ROGER CAYZER

Since retiring as the National Trust Regional Building Manager for the East of England Roger is now one of the 62,000 National Trust volunteers. In this role he is using his experience of a long career in built heritage to support the cause of conservation and heritage in the UK. This includes the Traditional Building Skills Bursary Scheme where apprentices are offered a recognised training scheme to develop craft skills with a conservation bias. With a continuing interest in raising the profile of conservation standards Roger attends training initiatives within the National Trust, external organisations and gave a presentation at the 2014 International Planned Preventative Conservation Conference in Milan. With a long career in built heritage he worked for many years at the Palace of Westminster as a project manager working on a portfolio of government buildings throughout Westminster and Whitehall. Prior to this he was employed as a building surveyor for the Department of Architecture of the Greater London Council responsible for the care of predominantly historic buildings.

TOM NISBET

Tom Nisbet has worked as a Building Surveyor for the National Trust for over 10 years, providing professional building and conservation advice, building surveying and project management services to National Trust properties in London and the South East. A keen advocate of professional development, Tom leads the programme for the RICS Assessment of Professional Competence within the Trust. He is also a member of the Editorial Advisory board for the *Journal of Architectural Conservation*. His qualifications include an MSc in Conservation of the Historic Environment and membership of the Royal Institution of Chartered Surveyors.

Acknowledgements

CHAPTER 1 BUILDING RECORDING

SADDLESCOMBE FARM
Bob Edwards
Forum Heritage Services
Room 5, Dorset House
No. 1 Church Street
Wimborne
Dorset BH21 1JH
www.forumheritage.com

EMLEY FARM
David Martin and Barbara Martin
Archaeology South-East
Units 1 and 2
2 Chapel Place
Portslade
East Sussex BN41 1DR
www.archaeologyse.co.uk

CHAPTER 2 MASONRY

HARDWICK HALL
Rodney Melville and Partners
Architects and Historic Building Consultants
10 Euston Place
Leamington Spa
Warwickshire CV32 4LJ
www.rodneymelvilleandpartners.com

THORINGTON HALL
Pick Everard
Weaver House
9 Looms Lane
Bury St Edmunds
Suffolk IP33 1HE
www.pickeverard.co.uk

CHAPTER 3 TIMBER

BROCKHAMPTON GATEHOUSE
Ian Stainburn
Caroe and Partners Architects
ABE Estate
Bromyard Road
Ledbury
Herefordshire HR8 1LG
www.caroe.co.uk/

SADDLESCOMBE DONKEY WHEEL
Peter McCurdy
McCurdy & Co. Ltd.
Manor Farm
Stanford Dingley
Reading
Berkshire RG7 6LS
www.mccurdyco.com

CHAPTER 4 STONE ROOF COVERINGS

THE FLEECE INN
John Goom
John Goom Architects
108 High Street
Evesham
Worcestershire WR11 4EJ
www.johncgoom.co.uk

ACKNOWLEDGEMENTS

BELTON BOAT HOUSE
Nick Cox
Nick Cox Architects
77 Heyford Park
Upper Heyford
Oxfordshire OX25 5HD
www.nickcoxarchitects.co.uk

LYTES CARY MANOR
Sam Wheeler
Philip Hughes Associates
Old Manor Stables
Tout Hill
Wincanton
Somerset BA9 9DL
www.pha-building-conservation.co.uk

CHAPTER 5 THATCH

BERG COTTAGE
Keith Quantrill
25 Dunstable Road
Flitwick
Bedfordshire MK45 1HP
www.thatchconsultant.co.uk

SELWORTHY COTTAGE
John Letts
Historic Thatch Management Ltd
Castle Lodge
Taunton
Somerset TA1 4AD
jbletts@btinternet.com

Glossary

Term	Explanation
arris	Sharp edge at the junction of two surfaces. Usually used to refer to the edge of a brick or stone
butts-up ridge	The most common way of ridging a roof in the West Country; combed wheat reed (reed straw) is laid 'butts' up (i.e. ears down) along both sides of the ridge and fixed with external hazel 'liggers' (2–3 cm diameter rods) and twisted hazel spars; the butt ends meet along the crest of the ridge and are pushed together tightly and fixed in order to create a watertight junction; the ridge protects the fixings of the final course of the main coat
CA	Contract administrator
chamfer	A bevelled edge on a board formed by removing the sharp corner. Generally used on mouldings, edges of drawer fronts, and cabinet doors
coat work	New thatch applied to a roof, in contrast to the old weathered surface (and underlying layers of old thatch including the original base coat)
CPI	Co-ordinated project information
draw boring	Technique using dowels to reinforce mortise and tenon joinery
dreft	A tool with a flat, corrugated surface that is used to dress (tap) wheat reed (reed straw) or water reed into position after fixing in order to tighten the coat and create a uniform surface
flaunch	Weathering to the head of a chimney stack. Usually formed of a mortar bed providing support to pots
fletton	Relatively soft and porous brick made from Oxford clay, of which a large amount comes from near Fletton in Cambridgeshire
flitch plate	Flat metal plate used in construction of flitched beam (i.e. beam built up of timber pieces and flitch plate bolted together)
frog (bricks)	Indentation into top surface of brick, formed by kick on stockboard against which clay is thrown, to push clay into corners of mould to form sharp arrises
inalienable	When related to property, 'inalienable' means that a property cannot be compulsorily purchased or transferred against the National Trust's wishes unless through special parliamentary procedure. It is an increased level of protection which only the National Trust and the National Trust for Scotland are able to apply, and they will do so to places of exceptional natural or historic value
latence	Potential mortar defects which may show only once mortar dries
leaf (bricks)	Brickwork is said to be one leaf thick if it has a total width equal to the length of one of its regular component bricks

Continued

Term	Explanation
leggett	Bat-shaped tool with grooved surface used for dressing thatch (also known as bat or dresser)
ligger	Round length of hazel or willow from saplings, usually approx. 5 ft long. Used for securing thatch and for decoration. Secured using spars
nogging	Horizontal timber member used between the studs of a framed partition
pargetting/pargetted	Decorative relief or indented markings to render surfaces. Commonly found in vernacular architecture in eastern counties of England
pintle	A pin or bolt, especially one on which something turns, as a hinge
pozzolan	Materials added to lime mortars to enable them to set more rapidly, e.g. ash or brick dust
quarry/quarries (glazing)	A square or lozenge-shaped piece of glass used in leaded casement windows
quoin	Dressed brickwork or stonework at the corner of a building
raking	To be set at sloping angle
soakers	Small flexible metal flashings laid with tiles at junction with abutment to ensure water remains on tiled surface. Usually concealed beneath cover flashing
spar	Split hazel or willow saplings about 2 ft long, twisted to form fixing to secure new thatch to old or to secure liggers
spere	Screen across the lower end of a medieval hall, concealing the screens passage between the hall and the kitchen
spere-truss	Structure rising from trusses fixed to the side walls of a timber-framed hall which coincided with the position of the screens passage
tilting fillet	Angled fillet fixed to upper surface of rafters at eaves to provide adequate support to lowest course of tiling in order to maintain angle of tile surface and tight junction with course above
voussoir	An individual wedge-shaped cut brick or stone forming part of an arch or vault
withes	Tough flexible branch of an osier or other willow, used for tying or binding
wrap-over ridge	Where ridge thatching material is centred on roof apex and extends evenly down both roof slopes. As opposed to butt-up ridge detail where ridge is formed of two lengths with a joint at the apex
yealm	A prepared layer of wet straw or sedge, approximately 18 in wide and 4 in thick, for use in coat work or ridge saddles

PART 1

Conservation philosophy and principles

The philosophy and principles encompass the reasons for the work and desired outcome. The philosophy is the sum of the factors which drive the need for undertaking building works; these will be predominantly conservation based for historic buildings, but might include physical, commercial, environmental or other drivers in different works. The principles embedded within a project set out the need for appropriate research and ensure that significance is recognised, understood and respected.

CONSERVATION PHILOSOPHY

A clearly defined brief needs to be drawn up prior to writing a specification, following survey and investigative works. This will describe the *conservation philosophy* being adopted, whether repair, maintenance, conservation, restoration and/or adaptation.

- *Repair*: work beyond the scope of maintenance to remedy defects caused by decay, damage or use. Includes minor adaptation to achieve a sustainable outcome, but does not involve restoration or alteration.
- *Maintenance*: routine work which is often necessary to keep the building fabric in good order.
- *Conservation*: the process of managing change to a significant place and its setting in ways that will best sustain its heritage values, whilst recognising opportunities to reveal or reinforce those values for present and future generations.
- *Restoration*: returning a site to a known earlier state, on the basis of compelling evidence, without conjecture.
- *Adaptation*: use of appropriate additions and/or alterations to secure the future use and viability of a historic building.

The philosophy should provide a clear framework demonstrating how decisions about physical interventions in the fabric of the building will be made, and the basis on which these should be undertaken.

The introduction to each of the case studies sets out the context for the work by describing the conservation philosophy adopted. This helps to ensure an appropriate and sensitive approach which preserves the historic fabric, enabling repair and conservation rather than restoration.

CONSERVATION PRINCIPLES OF HERITAGE ORGANISATIONS

The specific driver for any works to historic or vernacular buildings, irrespective of listed status or function, should be the significance of the fabric itself.

The National Trust has maintenance strategies and policies to protect and enhance the physical estate. To the Trust, 'conservation' is the careful management of change. It is about revealing and sharing the significance of places and ensuring that their special qualities are protected, enhanced, enjoyed and understood by present and future generations. The driving force behind how it manages this is set out in these conservation principles.

- *Significance*: we will ensure that all decisions are informed by an appropriate level of understanding of the significance and 'spirit of place' of each of our properties, and why we and others value them.
- *Integration*: we will take an integrated approach to the conservation of natural and cultural heritage, reconciling the full spectrum of interests involved.
- *Change*: we will anticipate and work with change that affects our conservation interests, embracing, accommodating or adapting where appropriate, and mitigating, preventing or opposing where there is a potential adverse impact.
- *Access and engagement*: we will conserve natural and cultural heritage to enable sustainable access and engagement for the benefit of society, gaining the support of the widest range of people by promoting understanding, enjoyment and participation in our work.
- *Skills and partnership*: we will develop our skills and experience in partnership with others to promote and improve the conservation of natural and cultural heritage now and for the future.
- *Accountability*: we will be transparent and accountable by recording our decisions and sharing knowledge to enable the best conservation decisions to be taken both today and by future generations.

The Society for the Protection of Ancient Buildings (SPAB) was founded by William Morris and other notable members of the Pre-Raphaelite Brotherhood in 1877 to oppose what they saw as insensitive renovation of ancient buildings in Victorian England. Today the SPAB still operates according to Morris's original manifesto. Its principles are:

> *Repair not restore:* although no building can withstand decay, neglect and depredation entirely, neither can aesthetic judgement nor archaeological proof justify the reproduction of worn or missing parts. Only as a practical expedient on a small scale can a case for restoration be argued.

Under the Planning Acts, the SPAB must be notified of all applications in England and Wales to demolish in whole or part any listed building; this includes ecclesiastical buildings which are normally outside the planning process.

The Burra Charter defines the basic principles and procedures to be followed in the conservation of Australian places. As the Charter observes: 'conservation is based on respect for existing fabric, use, associations and meanings'. It also emphasises the importance of the fabric in understanding the site:

> The traces of additions, alterations and earlier treatments to the fabric of a place are evidence of its history and uses which may be part of its significance ... the cultural significance of a place is embodied in its fabric, its setting and its contents' – that is, the 'historic building itself is an artefact'.

The International Council on Monuments and Sites (ICOMOS) defines its conservation principles as:

Imbued with a message from the past, the historic monuments of generations of people remain to the present day as living witnesses of their age-old traditions. People are becoming more and more conscious of the unity of human values and regard ancient monuments as a common heritage. The common responsibility to safeguard them for future generations is recognized. It is our duty to hand them on in the full richness of their authenticity. (*International Charter for the Conservation and Restoration of Monuments and Sites*, 'The Venice Charter', 1964)

Historic England is the government's statutory adviser on England's historic environment. It is responsible for assessing buildings for listing and puts forward its recommendations to the Secretary of State for Culture, Media and Sport who makes the final decision. The actual management of listed buildings consent applications and approval of these is the responsibility of local planning authorities.

Historic England's six key conservation principles are as follows.

1. The historic environment is a shared resource.
2. Everyone should be able to participate in sustaining the historic environment.
3. Understanding the significance of places is vital.
4. Significant places should be managed to sustain their values.
5. Decisions about change must be reasonable, transparent and consistent.
6. Documenting and learning from decisions is essential.

A charity, the English Heritage Trust is licenced to care for and open to the public the National Heritage Collection of more than 400 state-owned historic sites and monuments across England. There is some overlap with the Trust's portfolio, but they do not have the farmed estates or tenanted properties the Trust has responsibility for .

Equivalent organisations to Historic England and the English Heritage Trust in the other home countries are:

- Cadw in Wales
- Historic Scotland
- The Northern Ireland Environment Agency.

Consideration of conservation principles

Conservation principles should be set out to ensure that conservation work to a historic building shows the greatest respect for its historic fabric and overall cultural heritage value. It should be achieved with minimum intervention and involve the least possible loss of historic fabric.

Preventive conservation, in the form of regular monitoring and maintenance, is the key tool in conserving historic properties. Early identification of deterioration enables it to be slowed or stopped, thus avoiding unnecessary repairs. Regular maintenance is preferable to repair, and timely and effective repair is preferable to restoration. It follows in conservation 'best practice' that when repairs, additions and alterations are made to historic buildings and structures, these should be handled 'truthfully' to make clear what is original fabric and what is new (introduced) fabric.

For example, in carrying out repointing to brickwork, ageing or distressing the new mortar should normally be avoided in preference to allowing natural processes to blend the new work with the existing over time. Modern analysis can be used to establish the mix and colour of sand to match existing materials.

An important consideration when the works include alterations is 'reversibility', i.e. that they are capable of being reversed so that the site can return to its previous state.

From a financial perspective, finding a new use may be the only effective way of retaining a building. The basis of judging the most appropriate alternative use is likely to relate to least interference with the fabric, and therefore only those uses which reduce or avoid interference should be acceptable. For example, the sympathetic reuse of a disused historic farm building or barn for light industry could be achieved by inserting a whole new 'stand-alone' structure, which requires minimal connection, within the historic farm building 'shell'. In this way, hygienic linings could be applied to the new structure without affecting original surfaces and minimal fixings to the original structure. These new insertions could be easily reversed in the future. Conversely, adaptation of that building to a residential dwelling would usually be a last resort as it would typically require significant intrusive alterations to the historic structure, which could be difficult to reverse.

Buildings should have a viable use to ensure their environmental and financial sustainability. The issue of reducing energy is an ethical matter related to the greater social good of reducing impacts on the environment. This is a careful balance between environmental impact and the conservation and social value of the built cultural heritage.

The issue of the embodied energy in the existing building and its energy consumption should also be considered. Embodied energy is the energy and carbon consumed on manufacturing of building materials and building construction, including the transport in connection with these activities. It is therefore essential that there is a detailed understanding of the actual energy performance of the building and, irrespective of statutory compliance, that there is an acceptance of the building's limitations and energy 'trade-offs'.

A key principle should be to understand what you have, how it works, what you need and how to use less energy more effectively. The materials used in pre-1919 buildings allowed the fabric to 'breathe', whereas modern materials insulate; failure to understand traditional building processes and their interaction with modern technologies could cause damp problems and irreversible damage. Before preparing specifications for work that includes energy efficiency measures, consider the existing energy performance of the building; for example, when specifying repairs to a window frame, detail sympathetic draught-proofing measures or secondary glazing.

Implementation of the principles

A thorough survey and understanding of a building's construction and the materials used is essential before undertaking any works. How comprehensive these are will depend upon the building's significance, usually stemming from its listed status, the works proposed and the time of year. Outcomes of surveys should include:

- a drawn, scaled record giving floor plans, elevations and cross-sections and/or long sections where needed
- a fully indexed digital photographic record to support the drawings
- an analysis and interpretation of the development of different periods of the historic building
- an analysis of the form and character of the building and where appropriate of its landscape character in the locality and setting
- identification of inappropriate materials used in previous maintenance that may affect the long-term structural integrity of the building
- a description of the building's construction and materials
- prioritisation of repairs and, where scaffolding is necessary, identification of all essential repairs so that this can be undertaken whilst the scaffold is in place.

CONSERVATION PLAN

The principles can be put into practical steps in a conservation plan. This is a holistic document, drawing together research which will lead to thorough understanding of the asset's significance. This will then inform any discussion concerning the building which will ensure appropriate decisions are taken.

The plan should be structured and include an array of information relating to the history and construction of the asset. This is achieved through detailed building recording. The document also explores the cultural significance of the asset. It may also contain the long-term vision and detail regimes by which it can be maintained or conserved. The conservation plan will present valuable information to those responsible for the building's conservation or alteration. It will also assist with the design and justification of any necessary works which require statutory consent, aiding the progress of an application.

HERITAGE LEGISLATION

Statutory legislation exists to protect the environment in which we live; to ensure that built structures are fit for purpose and to reduce the risks posed during construction and, later, to the occupants and those maintaining structure or services. These include planning and heritage legislation, building regulations and acts pertaining to the protection of health and safety of individuals, and most notably the Construction (Design and Management) Regulations. It is very important that all statutory consents and associated compliance issues are identified during the early planning stage of the project. This should reduce the likelihood of unnecessary time and cost over-runs to the project and design process.

The statutory legislation relating to heritage protection in England and Wales is the Planning (Listed Buildings and Conservation Areas) Act 1990; in Scotland, the statutory heritage consent process is governed by the Planning (Listed Buildings and Conservation Areas) (Scotland) Act 1997. The Act derives much of its philosophy from the ICOMOS Burra Charter which encourages those involved in the management of heritage assets to thoroughly assess and understand the *cultural significance* of a place. (According to the

Burra Charter, Article 1.2, 'Cultural significance means aesthetic, historic, scientific, social or spiritual value for past, present or future generations. Cultural significance is embodied in the place itself, its fabric, setting, use, associations, meanings, records, related places and related objects. Places may have a range of values for different individuals or groups' and (Article 1.4) 'Conservation means all the processes of looking after a place so as to retain its cultural significance'.) The principles applied from this will then, in turn, guide the successful management of the place and help to inform its conservation. The application of this legislation is further guided by the National Planning Policy Framework.

Listed buildings

This legislation covers buildings or structures which have been judged to be of special architectural or historic interest and have been included on a list approved by the Secretary of State. These 'listed' buildings obtain the protections set out in the Act as do 'any object or structure fixed to the building' or 'any object or structure within the curtilage of the building which, although not fixed to the building, forms part of the land and has done so since before 1st July 1948' (Planning (Listed Buildings and Conservation Areas) Act 1990, Part 1, Section 5). This will often include boundary walls, lodge houses and follies or garden structures.

The designation of listed buildings is subdivided as follows (as of 2014).

England and Wales		Northern Ireland		Scotland		Definition
Grade	% of total listing	Grade	% of total listing	Grade	% of total listing	
I	2.5%	A	2.4%	A	8%	Exceptional interest nationally
II*	5.5%	B+	4.8%	B	51%	Particularly important and of more than special interest
II	92%	B1/B2	92.8%	C(S)	41%	Of special interest, particularly regionally

While all provisions of the Act apply to each listed structure, the listing grade helps to inform decisions by the statutory bodies.

The legislation accepts that listed buildings form part of the nation's building stock and that alterations may be required to ensure the continuation of use and longevity. Consent is required for works to the listed structure which will impact upon its special architectural or historic interest. This can be obtained from the local planning authority through a formal submission process which will require the applicant to robustly justify the proposals. It is likely that they will be challenged to ensure that the works are both necessary and will be undertaken in the most appropriate way.

Conservation Areas

Under the same Act, designated Conservation Areas receive protection against works which will be detrimental to the character or appearance of the designated area. The application process for alterations is similar to that for listed buildings.

Scheduled Ancient Monuments

Monuments considered to be of national importance receive protection under the Ancient Monuments and Archaeological Areas Act 1979. Monuments are defined by Historic England as 'deliberately created structures, features and remains'. Once scheduled under the Act, they receive a very high level of protection as administered by Historic England on behalf of the Secretary of State. The majority of Scheduled Ancient Monuments are uninhabitable ruins or remains. By scheduling monuments under the Act, it is intended that the historical asset will receive preservation as a monument, as far as possible without alteration.

Planning permission

Applications for works under the Town and Country Planning Act 1990 will take into account and be influenced by the impact of the proposed works on the siting of protected historical assets, whether or not additional consent is required.

PROCUREMENT AND ITS RELATIONSHIP WITH SPECIFICATION

The procurement of building services can vary significantly from one project to another and is guided by conservation requirements of the building and particular needs of its owner. Procurement has an intimate relationship with the specification and the preliminaries, all of which must be set with the same objectives and priorities.

For the purpose of this volume, 'procurement' relates to the selection of building contractor and/or subcontractors and specialists, and the contractual relationship between them and the client or building owner for the purpose of delivery of services and/or works.

Procurement should look to address and understand three key areas in order to minimise risk.

- *Finance*
- *Quality* of the delivered scheme (risk to cultural significance through inappropriate design or risk of poor workmanship)
- *Time frame*, especially where urgent repairs are required

In all projects, the finished product is always a balance of these three elements and it is important to determine at the outset which of these considerations will take precedence should changes occur in the project. If time frame is the top priority then costs may increase and quality decrease; if it is cost then time frame will increase and quality decrease; and if quality is the primary requirement then cost and time frame may increase.

The case studies in Part 2 demonstrate a broad array of procurement options available to the conservation professional, each delivering a successful outcome. In some cases, alternative approaches may have delivered an equally successful solution, and in others the selection was more restricted. In each case, the solution has been tailored to the particular requirements and situation of the building or structure.

Common procurement routes

There are three principal construction procurement routes which can form the basis of the selection for conservation works for significant projects. Each has benefits which can be weighed up against the identified risks for a specific scheme. In their basic forms, they are the most widely recognised industry standards. Other bespoke options are available, deriving from them.

The subject of construction procurement is broad and many publications are available which provide detailed interpretation. The following is intended as a brief introduction only, and in relation to their use with conservation contracts.

Traditional

A widely recognised design-led route, where a client-appointed designer researches and designs works to be undertaken, obtains statutory consents and invites tenders based on a detailed works package, often in conjunction with an appointed cost consultant. This is a robust and well-tested linear structure which offers reasonable cost and design control due to the front-loaded design input and good quality control, derived from detailed drawings, pre-liminaries, specification or bill of quantities. The methodical progression through this design stage works well with the level of detail required for conservation works. Additional enabling works can be undertaken to help reduce the number of unknown factors and reduce the requirement for contingencies, both financial and time. It may also identify a requirement for specialist works which have a limited skills base requiring specific subcontractors to be included in the procurement documents.

The main limitation of the traditional route is the gestation period between initial design briefing and commencement of works on site. Some compromises can be reached by appointing a contractor sooner, by interview assessment or by tender with partially com-plete information (two-stage tendering). This can enable a more rapid start and allow the contractor's often valuable input into the design. This is particularly helpful where intricate craft skills are required. This is, however, balanced by reduced cost accountability.

This process can lead to future works of a similar nature being negotiated with a single contractor who has demonstrated, through the traditional procurement route, good quality of workmanship and good working relationships with the client and design teams.

Design and build

A single organisation is employed by the client to both design and complete works in accordance with a set of agreed criteria. This path is more suited to new-build construc-tion; its inherent inflexibility, lack of competitive or comparative tender, and delegation of design duties to a party who may be more motivated by minimising cost/time than ensuring specialist details all greatly increase the risk to the quality of the delivered scheme. In the majority of conservation projects, this risk is unacceptable and this route inappropriate.

Construction management

Contractors often hold a wealth of knowledge and experience which remains unused until works have been designed, tendered and commenced on site. They can be better placed than some of the conventional members of a design team to offer strategic advice on such issues as phasing or defining specific work packages. They will often have a greater depth of appreciation for the physical undertakings of the proposals, material availability and how specialist skills can be best utilised. This is especially pertinent with conservation contracts.

To tap into this resource, an individual from a contracting background can be appointed to sit on the design team in the role of managing contractor with the architect/designer, quantity surveyor/cost consultant and engineers from the conception of a scheme. Directly appointed by the client, he/she will assist with the formulation of the design during the design phase and be responsible for liaising with independently appointed subcontractors during the construction phase. This level of insight and control can be highly beneficial to an intricate conservation scheme, provided an individual with the appropriate knowledge and background can be identified. The greater breadth of knowledge on the design team enables early identification of issues to provide greater cost certainty. The additional cost of this appointment to the client will often be balanced by reduced contractor overheads.

This method does create some uncertainty with the overall cost of the project. Costs only become apparent as the various work packages are let, but it does allow the client to reduce or remove work packages if cost is a limiting factor.

Whichever method is adopted for the tendering process, it is important that a robust and well-documented procurement procedure is followed. This needs to be designed so that fairness and equality can be demonstrated firstly to those tendering and secondly for audit purposes. In the case of grant-funded projects (e.g. European or Heritage Lottery), these procedures are specified by the granting bodies and must be adhered to as a condition of the grant (see Procurement legislation below).

Partnering and negotiation

Partnering is a term often raised in connection with procurement. While not a procurement process in its own right, it is a philosophy which encourages a client and a contractor to work together to mutually beneficial outcomes. This is usually through the development of an ongoing working relationship, which alleviates the need for regular tendering, reducing costs and hastening time frames. Clearly, there is additional financial risk to such an arrangement, and it is one that requires both trust and careful management, usually aided by the use of a good cost consultant.

While this approach does not immediately appear to lend itself to conservation practices, as it will restrict the breadth of expertise across different projects with bespoke requirements, it does have some merit. In a number of the case studies in this volume, including the works to the donkey wheel at Saddlescombe Farm, a local specialist was identified who had unique experience of similar work or in-depth knowledge of local techniques. In circumstances where the financial risk can be mitigated and justified through the use of a building or cost consultant, it can be appropriate to negotiate directly with the individual and enable the benefit of their experience to be used. This approach is generally best suited to specialist works which predominantly fall under a single trade, such as carpentry, stonemasonry or decorative plasterwork.

Different types of contract

A number of standard forms of contract exist which can be used by a professional to provide assurance to both client and contractor by laying out the terms of the relationship between all parties. These include payment terms, processes for variations and agreed procedures should difficulties arise. Such contracts have generally been well tested in the courts and form a clear definition of responsibilities, helping to avoid ambiguity, misinterpretation and subsequent disputes.

The Joint Contracts Tribunal suite (JCT) and New Engineering Contract (NEC) forms are both widely recognised in the construction industry. Projects which include a high proportion of mechanical and electrical works or civil engineering might benefit from alternative forms of contract written more specifically for such operations; examples include those produced by the Institution of Civil Engineers or the 'Model Forms' of contract drawn up by the Institution of Mechanical Engineers together with the Institution of Engineering and Technology. The duties and responsibilities of the client and contractor as well as other parties involved in the execution of these common forms of contract are well understood and therefore do not attract a surcharge to tenders, which may be a consequence of using less well-recognised contracts.

Procurement and the commercial challenge

The historic built environment is diverse. Consequently, the work that is required is equally diverse, often with commercial pressures including budget constraints, stringent deadlines and functional requirements. In the case of the sympathetic conversion of an otherwise redundant building, success against these criteria can ultimately determine a building's long-term viability and help to secure its survival in the modern world.

Careful identification of the risks and liabilities at an early stage is essential to provide a clear picture to the building owner or client before commitments are made. Consultation with suitably qualified building professionals should identify potential cost risks, the limitations for some of the specialist materials being used (e.g. use of lime externally during winter months) and guidance on implications against time frame or statutory requirements. A good appreciation and understanding of these issues enable the development of a scheme within realistic expectations before works are committed.

Whenever works are undertaken on a historic building, there is a risk of encountering unplanned works during the construction phase. It is therefore advisable to anticipate and allow a contingency sum for this. If there is time and available access, exploratory opening up should be undertaken to high-risk areas to mitigate this. The greater cost certainty that this brings can enable the contingency sum to be reduced.

An appreciation of the requirements of specialist trade skills and materials that may be required on a conservation project will allow detailed planning, enabling appropriate time frames and avoiding unnecessary expense. For example, where external limework is required, foresight can enable this to be undertaken during the warmer months and reduce the need for special protection.

Procurement legislation

European Union legislation requires that, where public sector funds are to be used on a scheme, tenders over a determined financial threshold are required to be advertised through the *Official Journal of the European Union*. This will include works funded through the Heritage Lottery Fund. The process requires additional administration to ensure fair and accountable procedure in line with European directives. Those unfamiliar with this process are advised to identify whether publicly funded works meet the threshold at an early stage to enable appropriate planning.

Contractor identification and selection

Building conservation 'best practice' promotes careful design, attention to detail (precision) and the use of carefully selected materials and components in the preparation of conservation specification. However, of equal importance to the quality and durability of the works is the competency of the contractors physically undertaking them. Identifying a contractor who has an appreciation for the attention needed and access to the specialist craft skills required is essential.

In order to maintain traditional skills, consideration should be given to a clause covering the provision of apprentices, with adequate weighting given to the decision during the tender process.

The diversity of vernacular buildings throughout the UK requires different skills and techniques in different locations. Consequently, local contractors will often have access to greater resources of trusted tradesmen with skills aligned to the local building techniques. Employing local stonemasons can be invaluable for their knowledge and experience; for example, their familiarity with the local stone to be used, where to source it, its strengths and weaknesses and best treatment.

Good contractors are often best sourced through recommendation. The professionals involved in the design team will usually have experience of different contractors and will be able to advise on firms of a suitable size for the proposed scheme. It can also be fruitful to develop relationships with owners of similar buildings to share experiences.

Once identified, it is beneficial to arrange a visit to a working site of a potential tenderer to gain an appreciation for the quality of site management, motivation of tradesmen and approach to unforeseen challenges and variations. It should be possible to gauge their experience in areas applicable to your own scheme. It is also advisable to have a telephone conversation with a number of their previous clients.

Some local contractors may not have the resources to manage a large project. Where this is the case and larger regional or national firms are required, a good specification can lead them to seek and use good local trade skills. When employing a large firm, try to identify the individuals you will be dealing with through the duration of the scheme and target work they have been involved with when looking for referees or site visits. Site management and contractor supervision are critical to executing the design and specifications and realising the objectives of the building project. To this end, the suitability of the site foreman and his/her relationship with the project team are of the utmost importance.

Some very specialist works, such as the conservation of a decorative plastered ceiling, may call for the use of a specialist from further afield. Such individuals can be identified in a similar manner outlined above. Where this work forms part of a larger scheme, the specialist or a selection of suitable candidates for this specialist task may be chosen by the client or design team and nominated as part of the tender procedure. This ensures the right specialist is undertaking the work.

Over time, contractors and an understanding of their individual strengths will become known to a building professional or building owner. However, contractors change and some are more adept at encouraging the skills of younger tradesmen/women and maintaining their quality of output than others. Being receptive to advice and identifying opportunities to meet new contractors on an ongoing basis is highly recommended.

Once suitable contractors have been identified, the work can be tendered. A tender can assess the cost, quality and timing of the work to obtain a true and comparative 'value' to determine the best appointment.

All tendering parties need to be made aware of the basis on which the assessments will be made at the point of tender and be presented with a clearly defined set of criteria and weightings. This will help to reduce the subjectivity and make the process more accountable. It would also be good practice to ask contractors to identify completed works of a similar nature as part of their assessment. This process can be undertaken in advance of the full tendering procedure, via a Pre-Qualification Questionnaire (PQQ), which will help to further sift potential contractors before putting them through the costly process of tendering. In each case, the information required will vary, focusing on the key challenges and constraints identified for the particular scheme.

Consultant identification and appointment

The identification of good consultants – designers, cost consultants, engineers or others – has many parallels with the identification of contractors. Identification of the individuals with whom you will have direct contact, their experience, skills and ability to adapt to your requirements are all of importance. Visiting previous schemes and discussing their work with past and existing clients are all valuable research and will pay dividends for the time commitment required.

The work can be tendered and costs will vary. Many consultants will, unless specified otherwise, tender on a percentage basis (a percentage of the final cost of the project), but you could request a fixed tender sum. Consultants such as architects also have suggested scales for fees or fee calculators which can give some guidance. As with contractor identification, it is important to consider all factors when choosing a consultant, from 'value' to experience, knowledge, approach, personality and way of working.

The approach to procurement will vary from project to project according to the type and scale of works, availability of skilled labour and individual site or project constraints, including required time frames and availability of funding. There is no right and wrong and a good consultant will help guide the process to a suitable outcome. The case studies in this volume will, we hope, help to identify some of the benefits and constraints of these techniques.

WRITING A SPECIFICATION

The specification forms a fundamental part of the procurement process. It is a critical tool in the management of any construction work. For conservation-based works where often the materials and workmanship being employed are more specialised, these factors are even more significant.

The purpose of a specification is to describe the works in qualitative and quantitative terms, so that they can be executed in exactly the manner intended, properly priced, cost controlled and accounted for.

This section provides a very general overview of specification structures and methods, and their application in conservation-based work, setting the context for the following case studies. The following is intended as a brief introduction only.

Structure

The structure and scope of specifications can vary significantly, depending upon the nature of a scheme and the parties involved in its preparation and use. Fundamentally, the detail, breadth, complexity and size of a specification needs to reflect the works for which it is intended, i.e. the monetary value, the significance of the asset and the risks involved.

Although the formats of specifications vary, they generally consist of four elements of information, as follows.

Preliminaries

These describe the conditions under which the works will be undertaken and may include specific employee site rules, commercial and time requirements. If a form of contract is to be used, they will usually state the prompts for completion. Any requirements for provisional sums, contingency sums and schedule of rates may also be included.

Materials and workmanship

This section provides the general technical requirements for materials and their use or application. This is typically trade specific.

Schedule of Work

The Schedule of Work will contain a detailed description of the work and may include schedules and drawings. Typically, the Schedule of Work either relates to particular locations/rooms or is trade specific. For larger projects, this element may be expressed in the form of a Bill of Quantities.

In the UK, the Common Arrangement of Work Sections (CAWS arrangement) is frequently used in the industry as a standard format for specifications and Bills of Quantities for building projects. It consists of a set of detailed work section definitions, all within a classification framework of groups and subgroups (for example, A10 Project Particulars; F10 Brick/Block Walling, and so on).

Drawings
These are typically floor plans, elevations and construction details, to provide clarity/
reference to the main text in the specification.

Format and methods
Techniques for writing specifications can vary significantly, and will usually depend upon
the specifier's background and experience. Generically the formats of specifications can be
summarised as follows.

National master specifications
National Building Specification (NBS) and National Engineering Specification (NES)
provide a pre-defined format which follows the CAWS arrangement, providing good
co-ordination between documents and drawings and enabling easy and quick access to
pertinent information.

The NBS and NES specifications draw on national experience and best practice to provide
a consistent format which is maintained to current standards. Their use by many building professionals of different disciplines can aid the effective transfer of information from
designer to quantity surveyor to contractor, promoting co-ordinated project information,
and is standard practice in many firms who are reassured by the detail and accuracy of the
standard clauses, which have often been tested in disputes. While office/practice expertise
and knowledge can be added, the temptation is often to pick and choose from pre-existing
clauses which can lead to a less bespoke product. In a large-scale new-build scheme, this
might lead to positive outcomes enabling more effective procurement and economies of
scale. In conservation work, however, where work can often be of much smaller scale and
require much more intricacy, this approach can feel cumbersome and result in a reduced
level of care and the inclusion of inappropriate methods or materials, both from the designer,
who is unlikely to be as intimately familiar with generic clauses as ones written first hand,
and the contractor who may not pick up on detailed modifications. The length and quantity of standard clauses can lead to unwieldy documents for small intricate projects, which
can drown out the detail of a specialist conservation contract if not used in a careful and
conservative manner.

Office/practice masters specifications
These are in-house developed specifications, often based on office expertise/knowledge
and will often be based around the CAWS arrangement, though may not be as closely
aligned as NBS/NES. Within a specialist conservation practice, or a small firm where all
members are familiar with the content and use of the company documents, they can be
very effective.

The reliance on individual knowledge and use will lead to inconsistencies between practices. With the pace of changing legislation, they can also quickly become out of date if not
diligently maintained. If used well, in conjunction with more bespoke clauses, they can be a
valuable tool to a conservation professional.

Manufacturer/supplier specifications

Manufacturer-developed specifications exist usually for stand-alone elements; these may also follow the CAWS arrangement and are often linked to the NBS.

This method of specification may offer specialist manufacturer/supplier expertise/ knowledge and will represent best manufacturer's recommendations and guidance. Where the accurate instruction is required for the correct use of a product, such as the application of a specialist coating or surface treatment, the inclusion of text direct from the manufacturer will help to avoid misinterpretation or ambiguity. Written by the manufacturer, they may stipulate the use of other products to their benefit. Left unamended, this can limit the options of a tendering contractor and result in higher costs.

Manufacturer/supplier-written specifications in conservation projects are rare as stand-alone specifications, and typically only provide component elements.

Bespoke specifications

Custom or project-specific specifications usually rely on individual expertise/knowledge. These are often used for small or very specialised works elements and, as a result, are unlikely to follow the CAWS arrangement.

This method of specification may offer the best specialist expertise/knowledge. They can be more flexible to unusual work and tailored to suit the task within the context of a conservation project. They can also take account of the needs of the associated disciplines involved on a particular scheme and adapted through partnership agreements with specialist contractors. They are inevitably time consuming to prepare and more costly as a result. Where funds and time allow, this approach will usually lead to the clearest transfer of information from designer to contractor and deliver the highest quality of result.

Bespoke specifications are often deemed essential for fine or specialised conservation work such as decorative plasterwork or gilding. Short bespoke sections may be referenced or included as subsections within NBS or office/practice masters specifications to get the benefit in a more economic and timely format.

Within these generic formats, there can also be variation in the different methods of specification. These can usually be summarised as 'prescriptive' (the process to be undertaken) and 'performance' (what is to be achieved) forms.

Method	Submethods	Examples	
		Component description (examples)	Reference
Prescriptive methods	Proprietary	Particular brand or model	Manufacturer's recommendations
	Descriptive	150 mm thick, green finish	British Standards
	Process	Four passes of a 5-tonne roller	British Standards
Performance methods	Performance	Two-hour fire-rated	British Standards BBA Certificates
	Price	Provisional sum	-

In the context of conservation, all of these formats and methods are applicable but their application will depend upon the specifier's approach as well as the nature of the work.

Implications

Conservation works are generally undertaken within the spectrum of the construction industry, whose collective experience is usually focused on new-build and refurbishment type work. If conservation specifications are not categorically clear in their requirements, then common practice will usually prevail and this may not be appropriate.

Specification for construction work is normally product led, with specifiers not being cognisant with the site practice element; specifications are usually 'silent' on the process aspect. In conservation work, however, the specifier could be the expert in the process (for example, with lime), and the specification may therefore have a greater emphasis on product and process.

There is generally a lack of standards (British Standards, codes of practice, etc.) specific to conservation work. As such, there is an onus on the specifier to be clear on the standards for quality, method and principles, rather than cross-referencing with existing standards.

Specification for construction work often stipulates exactly what is to be done. However, with conservation work, it can be equally important to specify what should not be done.

PART 2

Case studies

Chapter 1

Building recording

Understanding a historic building is the first step towards decisions about management, repair and alteration. Conservation projects should always begin with building recording, but it is often overlooked despite the fact that it usually represents a small percentage of the total cost in both time and money. By identifying architectural, historic and archaeological features, the analysis of fabric and structure contributes to successful conservation and carefully balances the significance of the building, the owner's requirements and appropriate techniques.

To plan any building work, you need to gain a full understanding of the structure and its environment by conducting surveys encompassing the physical remains, their context and the natural environment. The National Trust has developed a generic brief to ensure specialist consultants undertake a historic building and structures survey (HBSS) fully, on a consistent basis and to a high standard. A copy of this brief is included in the appendices, along with a copy of the template used. As well as being a record of the fabric and structure, this includes an analysis and interpretation of a building's origins, how it was used, and the processes involved in its development.

Building recording is only one aspect; you would also need to look at hazardous substances, such as asbestos, and protected species, such as bats. The initial survey may lead to mitigation which could range from a full fabric survey to a watching brief to changes to plans or time frame. The range of specialisms required for a HBSS might involve new technology such as dendrochronology, laser scans, 3D modelling and more. The scope for building recording is very broad in terms of focus, emphasis and level of detail. It is essential to state precisely what the deliverables are in a good recording specification, as they would be in a trade specification, to give the contractor confidence about what is expected in a particular recording project. This range of possible deliverables is evident when comparing the two Trust properties that provide the case studies for this chapter: Saddlescombe Farm is a large farm complex, normally open to the general public, whereas Emley Farm is a single structure used as a holiday let.

The report needs to be for internal use, but because it is effectively a historic record, it could also be used for academic purposes. It therefore will need to be of a professional standard and appropriate consideration will need to be given to archiving and distribution.

For each building surveyed, the completed report should provide the most recent digital record, capable of being added to or enhanced as the need may arise. It is an essential tool in the planning processes. For the Trust, as for many organisations, the report will be incorporated into a historic buildings, sites and monuments record and copied to local authority heritage, environment and planning services.

Beyond its own guidelines, the Trust follows Historic England building recording principles outlined in *Understanding Historic Buildings: a guide to good recording practice*. This latest Historic England guidance document gives full details of the most general applicable building recording techniques and includes a discussion of appropriate levels of survey and a set of drawing conventions. It outlines the latest technology for building surveying, which is a fast-changing area.

The discipline of building recording has a number of different titles. Both of the case study specifications use the very common title of vernacular building survey, but this may be called a vernacular building assessment or, as is more currently popular, interpretive historic building survey.

There are also a variety of reports resulting from the recording. In the first case study, Saddlescombe Farm, an archaeological and historical assessment and conservation statement is specified to help the Trust more fully assess its historical significance, develop a long-term vision for the property, inform further work and guide management decisions. As with many Trust properties, there had been previous studies, recordings and surveys done at Saddlescombe. While it is important not to duplicate work that is adequate and sufficiently accurate, the recording specified here does enhance and build on previous studies as seen by the report's flagging of additional buildings for potential listing and identification of new locations with possibilities for below-ground archaeology. In addition, the report specified here requests a full photographic record in order to provide a new monitoring point to the photographic record made a decade earlier.

The second case study, Emley Farm, is a Trust holiday cottage. The specification required a common and important end-product: an archaeological watching brief during the repair and refurbishment. Periods of repair often provide unique opportunities for opening up spaces that are otherwise inaccessible.

For an organisation such as the Trust, historic building records must provide information across a broad spectrum of users. As well as providing specialised and in-depth analysis to professional and academic (historical research) users, some end-products must be accessible to lay users. The report specified for Emley formed the basis for the guest guidebook for the house.

Saddlescombe Farm, Newtimber, West Sussex

Figure 1.1 Saddlescombe Farm. © Forum Heritage Services.

CONTEXT

Saddlescombe Farm lies in the South Downs and forms part of the Devil's Dyke Estate. The nucleated farm consists of a number of farm buildings and dwellings ranging in date from early 17th to late 20th century, indicating a gradual farm development rather than any single planned phase. The buildings cluster around three main yard areas, one of which, to the north of the farmhouse, was subdivided in the 19th century and is now predominantly a thoroughfare for walkers and gives access to the farm and cottages. The land surrounding the farm amounts to 243 ha (600 acres) and is of considerable archaeological and historic landscape value.

The farmstead has a documented history as a manor as far back as the Domesday Book. It was held by a number of different people and organisations in its history, including the Knights Templar in the 14th century. The Trust acquired it in 1995 from Brighton Borough Council, which had held it since 1925.

The farm buildings at Saddlescombe create a large 'regular multi-yard' farmstead where a number of usually adjoining yard areas have been created, typically to provide for cattle management. Recent analysis of farm plans in Sussex has mapped their distribution and it is possible to identify Saddlescombe as one of the few farmsteads in the South Downs where a regular multi-yard plan survives relatively unaltered compared to late 19th-century mapping.

Saddlescombe Farm began as a large mixed farm (sheep and corn). The importance of arable on the farm is clearly indicated by the presence of three aisled threshing barns, one of which was converted to provide a large granary. In the 19th century, cattle and dairying increased in importance at Saddlescombe as on many other downland farms. This resulted in

Figure 1.2 Saddlescombe Farm circa 1850. © Martin Frost.

Figure 1.3 Entrance to poachers' gaol. © Forum Heritage Services.

Figure 1.4 Granary barn interior. © Forum Heritage Services.

Figure 1.5 Stable interior. © Forum Heritage Services.

several ranges of buildings for cattle and the conversion of part of one of the barns for them. As a large isolated farmstead, Saddlescombe was required to be as self-sufficient as possible and so ancillary buildings, such as a forge and carpenter's workshop, were provided where equipment could be repaired.

Figure 1.6 Interior hearth to the forge. © Forum Heritage Services.

Figure 1.7 Milking parlour and feed stores. © Forum Heritage Services.

Figure 1.8 Tudor kitchen fireplace. © Forum Heritage Services.

The relative lack of modern intrusion and alteration to the farmyard and buildings contributes substantially to the historic significance and special atmosphere at Saddlescombe Farm. Some buildings are used as Trust estate workshops and low-level storage. The farmhouse provides facilities for staff and accommodation for the head ranger. The farm is occasionally open to the public.

CONSERVATION PRINCIPLES APPLIED

While there had been a certain amount of low-profile repair and a ten-year rolling programme of maintenance work at Saddlescombe Farm since it was acquired by the Trust, there was no clear and agreed long-term vision for its treatment and use. The Trust's archaeologist prepared the following specification which was a request for an archaeological assessment and conservation statement. It was a brief for the essential preparatory assessment to gauge more accurately the farm's historic value and significance, to identify issues, to recommend options, conservation policies and actions to guide its long-term future management.

SPECIFICATION OF RECORDING FOR SADDLESCOMBE FARM

1. **Extent of archaeological assessment to date**

1.1 Following Trust acquisition, an archaeological survey was undertaken and an initial vernacular building survey (VBS) was added in 1996 (Mayes, 1996).

1.2   The VBS consisted of a measured survey (1:50) with some limited building sections and elevations collated through CAD. Written descriptions identified elements of historical importance and a photographic record showed the condition of the farm buildings in 1996. All of this was entered onto the Trust archaeological database – the Historic Buildings, Sites and Monuments Record (HBSMR), where buildings are inventoried according to their unique identifying numbers. A chronological history of the manor is also provided in the VBS report.

1.3   Subsequently, the farmhouse building was renovated to provide accommodation and office space for the head ranger, with works involving retiling and repairing the roof. Archaeological recording and interpretation were provided to assess the nature of the roof timbers, their construction, development and repair.

1.4   In 2004, the Trust prepared a ten-year phased programme of essential repair and maintenance (Martin, 2004) for the farm buildings with the intention of keeping them in a stable and sound condition, pending any future plans for change. An archaeological recording brief was advised to prevent or mitigate any damage or loss of significant historic fabric that might be caused through the repair process. The most recent of these repair projects was for the 'mix and mill shed' with a report for the monitoring now due from Archaeology South-East.

2.   Purpose, scope and extent of survey

2.1   An archaeological and historical assessment and conservation statement is now required to help the National Trust more fully assess its historical significance, develop a long-term vision for the property, inform further work and guide management decisions.

2.2   The written assessment report will need to deliver the following outcomes.

2.2.1   Significance

- Individual building analysis. This will comprise an evaluation of the current vernacular building record and an assessment and interpretation of the significance of each discrete building in the complex. It will include text descriptions and digital photographs (exterior and interior) and some measured survey (ground plan and elevations where need dictates).
- An analysis of the character, significance and relative survival of the whole farm building complex and its context within the South Downs with specific reference to the historic landscape character and to the English Heritage project to characterise the farm buildings of West Sussex.
- A description of the historic phasing and coloured phase plans to demonstrate this.
- A fully indexed black-and-white photographic survey of each building, showing both interior and exterior elevations and feature details. An initial record of this type was completed in 1995 following National Trust acquisition, and this updated survey and catalogue will provide a new monitoring point after that ten-year period.

2.2.2 Issues and management options

The report should make an attempt at identifying specific conservation issues for the long-term maintenance and care of the building complex. It should also identify options for long-term management based on the farmstead's significance and capacity for change.

2.2.3 Impacts and mitigation through management choice

Linked to this, the report should attempt to offer up indications of likely impacts that might arise from following these options and how these might be overcome through the adoption of alternative forms of future management.

3. Survey programme

3.1 With due reference to the archaeological contractor's existing commitments, the survey should be undertaken as soon as the project design and quotation have been approved and a purchase order delivered.

3.2 Access can be made available at any time through negotiation with the National Trust property manager and head ranger.

3.3 Assuming that the survey will be under way during the month of August, the draft report should be presented for discussion at a meeting with the Trust's archaeologist, curator, property manager and head ranger in early–mid September. The completed report should be delivered within one month of that meeting or at a date agreed at that time. The photographic survey may need to follow at a later date and be presented separately, but this can be subject to negotiation.

4. Report production and distribution

4.1 Six copies of the completed report will be produced in a bound, double-sided A4 format. A digital copy will also be submitted with each paper copy in a Word or RTF format, supplied on CD. These copies will be distributed as follows.

- Three bound copies and CDs for main office.
- One bound copy and CD for the property (with photographic record, to be archived on completion of the survey).
- One bound copy and CD for the regional office.
- One bound copy and CD for the local authority.

5. Archaeological project recording form

5.1 A completed record form will be submitted by the contractor to the assistant archaeologist on completion of the survey.

6. Copyright

6.1 All reports and archive material generated from the survey are the copyright of the National Trust, unless otherwise stated.

7. Health and safety

7.1 The contractor will be required to demonstrate that they hold public liability insurance to a value of not less than £2,000,000.

7.2 The contractor will take sole responsibility for observing all current legal requirements concerning their or their employees' health and safety. The contractor will supply to the National Trust a copy of their health and safety policy on submission of the bid.

7.3 The contractor will also be required to provide a copy of a risk assessment for the survey.

8. General terms

8.1 The project will be undertaken by the contractor acting on an independent basis. Staff working on the project will not be deemed employees of the National Trust. Tenders should reflect this fact, and more specifically the contractor will take sole responsibility for the payment of tax, National Insurance contributions, etc. If VAT is payable, this too should be indicated in the bid.

9. References

Martin, A. *Ten-Year Projects Programme for Saddlescombe Farm* (National Trust, 2004).

Mayes, I. Saddlescombe Farm Vernacular Building Survey (unpublished report to the National Trust, 1996).

Shorland-Ball, R. and Wilson, C. Saddlescombe Farm West Sussex: Report on the Agricultural Collection for the National Trust (unpublished report, 2002).

Other reports derived from the property are available for consultation from the head ranger, based at Saddlescombe Farm.

10. Enclosures

- National Trust boundary map (also sent digitally)
- Report on agricultural collection (Shorland-Ball and Wilson, 2002)
- Statement of significance, sketch map and plan of buildings
- Vernacular buildings survey (Mayes, 1996)
- A copy of the two-volume photographic record will be made available for consultation via the property (currently held by Trust archaeologist)
- Disk copy of recent sample National Trust vernacular building survey (Littlewood Farm) report and report template as model for style of report

Specification author: Bob Edwards, Forum Heritage Services

COMMENTS ON THE RESULT

Whilst there are several listed buildings within the farmstead, the survey identified a number of others that are highly significant and may be worthy of listing, including the piggery, the forge and the carpenter's workshop. The archaeological value of the farmstead and many of the individual buildings was found to be high; it was recommended that any repairs to the buildings should respect, record and, in some circumstances, replicate small details that allow the development of the structures to be understood.

The potential for below-ground archaeology within the farmyard was found to be limited, particularly on the western part of the site where the ground levels appear to have been reduced, but deeper deposits were found to exist on the eastern part of the farmstead. To the south, where earthworks indicate the presence of a once larger settlement, the archaeological potential was found to be high.

Analysis of the buildings suggested that there are a number that have the capacity to accommodate new uses without having a detrimental impact on their individual character or the farmstead as a whole. In relation to the Trust's future management of the site, although some options were identified, it was clear that there was a need for an overarching purpose and vision for Saddlescombe Farm and for feasibility studies on any potential alternative uses. These were set out briefly within the report but subsequently investigated in the Trust's vision statement for the Saddlescombe Estate.

The report concluded that: 'the exceptional range and survival of buildings and plan form at Saddlescombe indicates that this is a highly significant farmstead. It is essential that every effort is made to retain its character through adequate maintenance and repair and, where appropriate, find new uses for the buildings'.

This has assisted with the ongoing conservation and management of the historic farmstead. Several of the buildings have now been sensitively converted to alternative uses, including facilities for visitors.

Emley Farm, Bowlhead Green, near Witley, Surrey

CONTEXT

Emley Farm lies in a shallow bowl at the foot of Rutton Hill, within the Surrey Hills Area of Outstanding Natural Beauty and the Green Belt. The surrounding 51 ha (127 acres) include the clump at the top of Rutton Hill to the north and fields approaching both Witley Park to the north and the Trust's Hindhead Common to the west.

It comprises a sizeable farmhouse (Grade II), the earliest period of which dates from the 16th century, and a picturesque group of vernacular farm buildings that reflect the mixed farming traditions of arable and rich grazing associated with the Low Weald. Included are two threshing barns, a granary (Grade II), a cartshed, cattle pens and stables.

The farmhouse consists of four two-storeyed ranges positioned around a small central courtyard or light well and is designed with its principal elevation facing south towards the farm track. The main 'public' elevations of the house (i.e. the eastern end wall of the main range, the entire south front and the full length of the west elevation) are stone built, but the remainder of the structure is timber framed, mostly now infilled with brick. The roofs are tiled. Against the northern side of the central courtyard is a lean-to structure containing steps leading down into the cellarage. The whole of the courtyard is today taken up by a lead-lined water 'cistern' fed with water from the internal roof slopes.

Figure 1.9 Front elevation (17th-century) of Emley Farm, with original wooden chamfered windows and historic glass. © Archaeology South-East.

Figure 1.10 Combination of local stone, 17th-century mullioned windows and early 20th-century tile hanging on gable end. © Archaeology South-East.

Figure 1.11 Seventeenth-century front door, with lock and holes for bar. Large old keys also survive. Note worn brick step. © Archaeology South-East.

Figure 1.12 Good example of early domestic 17th-century staircase, enclosed but not yet a major feature in its own right. © Archaeology South-East.

Figure 1.13 Timber framing dates in part to the 16th century. © Archaeology South-East.

Figure 1.14 Mainly 17th-century window with original wooden chamfered mullion and frame. Reflections show distorting effect of old lead and old glass. © Archaeology South-East.

The farmhouse was considerably extended in the 17th century and carefully modified in the early 20th century in the spirit of the Arts and Crafts Movement. The farm buildings appear not to have been improved or touched by intensive activity for more than 70 years. Set within a historic landscape of small pasture fields and hay meadows, at the end of a winding track, the farm is an important and rare survivor of the late medieval rural Surrey vernacular.

CONSERVATION PRINCIPLES APPLIED

The Trust was bequeathed Emley Farm in 1992 by Dr Dennison. He had informed the Trust of his intention in 1987, and the Trust had advised him on the conservation of the landscape and farm buildings thereafter. The Trust deliberately deferred making a judgement on the future of the property, but recognising the need to maintain the atmosphere of the place and provide a sustainable future for the buildings, it declared the property inalienable in 2004.

The archaeological brief in the next section is for an enhanced VBS and historic landscape assessment. Archaeology South-East, the contracts division of the Centre for Applied Archaeology, University College London, was commissioned by the Trust to carry out the survey. The stables were surveyed in September 2006 and the remainder of the farm in 2008 as part of a targeted programme of works to upgrade the existing VBS records and inform the repair and alterations to the property as a whole, prior to its conversion to holiday accommodation.

SPECIFICATION OF RECORDING AND REPAIRS TO EMLEY FARM HOUSE

1. Background to the project

1.1 The project aims to carry out sufficient repair, refurbishment and updating of the house and its immediate environs to enable its use as a holiday cottage while maintaining the historic integrity of the farmhouse and associated farm buildings and their setting within its historic landscape.

1.2 Conservation repairs will be carried out on the structure with a 'light touch' to conserve where possible as much of the historic fabric as can realistically be achieved. Where necessary, repair or replacement of fittings and fabric is to be on a like-for-like basis, using locally sourced vernacular materials and skilled craftsmen.

1.3 Previous National Trust records of the farm and farmyard comprise a vernacular building assessment (Higgins, 1988) and a conservation plan (Sekers, 2004). There has been no dedicated archaeological assessment to date of the buildings or of the landscape.

2. Purpose, scope and extent of survey

2.1 To inform the repair and interpretation of the house and surrounding complex, an archaeological survey and assessment is required before work begins. This will check and upgrade the existing vernacular building survey (Higgins, 1988), providing a modern, digital record which can then be related to the refurbishment needs, taking account of impacts and offering considered mitigation where needed. The assessment will include the farmhouse and farm buildings set around the farmyard, and will also address the research and record of the historic landscape of the farm to give necessary context. It will include the checking of existing sources and both fieldwork and documentary research, as detailed below.

2.2 A watching brief during the repair project is also required to monitor and record change. This will include both structural work on the house and outbuildings and also any below-ground interventions (service trenches, etc.) within or in the close vicinity of the buildings.

2.3 As well as the normal survey report, in view of the proposed use of the building for holiday accommodation, a separate, accessible illustrated summary of the main points of historic and archaeological interest of the farmhouse and its buildings and landscape is required. This might then more easily be worked up into an information leaflet for Trust customers.

3. Requirements of survey

3.1 For each building, the purpose of the survey is to assess the construction and historic phasing, with the particular intention of establishing its significance both as a discrete individual structure and with regard to its relationships as part of a historic group complex.

3.2 The resulting report will provide an enhanced and revised vernacular building survey with a focus on interpretation and understanding. This will inform the nature of any future repairs and any conservation measures that might be needed for the historic fabric.

3.3 It will also provide a baseline digital record against which any future changes to each building or alteration to the management of the property may be measured and recorded.

3.4 The survey will be undertaken along the lines of a standard historic building interpretation survey. It will examine and check the validity of any existing analysis and add further detail as needed. It will consist of a room-by-room record of the main structure of each building, in both a text-based and graphic format. For each building, it will consist of the following.

- A drawn, scaled record to be produced on CAD to give floor plans, elevations and cross-sections (where appropriate) unless existing accurate digital plans can be incorporated. If non-digital existing plans are found to be accurate and capable of scanning and adaptation, then this approach should be adopted. Unnecessary duplication should be avoided.
- A fully indexed digital photographic record to support the drawings.
- An analysis of the architectural structure and historic phasing.
- Identification of significant historic fixtures and fittings but without duplication of material already recorded within the conservation plan (Sekkers, 2004).
- An analysis of current condition and risks.
- Checking of documentary records and an understanding of the history of the complex, the building in particular, and of its direct historic landscape setting. All sources to be fully referenced.
- A written report that should include the following.
  - Site details.
  - Synopsis and location.
  - Overview of each building.
  - Significance of each building.
  - Detailed architectural description of the historical phasing of each building.
  - Detailed description of the building as existing at the date of current survey. This should include the interior (room by room) and exterior of each building; a description of its relationship with others in the complex; a statement relating to its landscape impact or setting. Identification of issues and recommendations for management. This section should indicate those features that have been identified as crucial to the historic integrity of the building and that must not be lost, and any that might be less significant. This should be practical guidance that is capable of application (rather like an impact assessment).

- Copies of CAD plans.
- Account of the historic background to the site. On the basis of knowledge gained, this will provide an interpretation of the main building ('tell its story') and of how it relates to the curtilage and wider landscape that gives it context and meaning.
- Index to photographs.
- The report format will follow the style that has already been developed and agreed between the Trust's archaeologist and the archaeological contractor whilst working on previous commissions during 2006 and for which a report template has already been prepared.

4. Survey programme and project review arrangements

4.1 The start date of the repair project is 4 September 2006.

4.2 Access to the property for the archaeological assessment to precede the refurbishment work can be made throughout August. The building repair specification will need to be consulted once the initial archaeological building assessment has been made to ensure that proposed repair or alteration will not impact adversely on significant historic fabric.

4.3 Attendance on site to carry out the watching brief will need to be negotiated between the contracted archaeologist, project manager and building contractor.

4.4 A draft report will be submitted within one month of the final site visit, with a final report to be agreed and presented within a further two weeks.

5. Report procedure and distribution

5.1 Five copies of the final copy of the report will be produced in a bound, double-sided A4 format. Three digital copies will also be submitted in a Word or RTF format on CD.

5.2 Copies of the report will be distributed as follows.
- Three copies plus CD for main office.
- One copy plus CD for the regional office.
- One copy for the property or the rural surveyor.
- One copy plus CD for the local authority

6. Archaeological project recording form

6.1 A completed form will be submitted by the contractor with the final copies of reports to the archaeologist at Heelis.

7. Copyright

7.1 All reports and archive material generated from the survey are the copyright of the National Trust, unless otherwise stated.

8. Health and safety

8.1 The contractor will be required to demonstrate that they hold public liability insurance to the value of not less than £2,000,000.

8.2 The contractor will take sole responsibility for observing all current legal requirements concerning their or their employees' health and safety. The contractor will supply the National Trust with a copy of their health and safety policy on submission of their quotation.

8.3 The contractor will also be required to provide a copy of a risk assessment for the survey.

9. References

Higgins M. National Trust Architectural Cards and VBS Emley Farm, Surrey (unpublished, National Trust, 1988).
Sekers, D. Emley Farm Conservation Plan (unpublished, National Trust, 2004).
National Trust digital property boundary map, supplied as PDF.
Specification author: David Martin and Barbara Martin, Archaeology South-East

COMMENTS ON THE RESULT

Two interpretive historic building surveys were produced, one for the stable building in 2006 and another for the main farmhouse in 2008. The aim of these surveys was to provide an overview of the date, sequence of construction and principal architectural features of the buildings. It was not regarded as a detailed archaeological record, nor taken as definitive.

The assessment involved a visual inspection of the fabric, both internally and externally, including any accessible roof voids. No intrusive techniques were carried out, as this would have been inappropriate and potentially damaging to the property. Interpretation of the fabric and fittings therefore relies principally upon inspection of the visible evidence. A measured survey was also undertaken. The inspection and resulting interpretation were supported by survey plans, digital photographs and written descriptions.

In addition to the site-specific inspection, a limited amount of documentary research was carried out in order to gain a general understanding of the historic context of the site. The research involved a brief examination of available primary records plus secondary sources and map evidence. This research was carried out at the Surrey History Centre.

Chapter 2

Masonry: brickwork and stonework

The case study specifications in this chapter show the diverse range of properties that the National Trust is responsible for. The first is Hardwick Hall, one of the best known and most prestigious properties in Trust ownership, open to public access most of the year with very high visitor numbers; the second is Thorington Hall, a former farmhouse, let to tenants; and the third is the Old Serpentine Works, Boswednack, a small granite-built structure, typical of the local Cornish style.

The specifications highlight a number of good conservation practices and each specifies sample panels showing the proposed masonry units and mortar characteristics. These are to be produced by the contractor and approved by the contract administrator prior to commencing the work. There are many ways of setting the standard for sample approval and Hardwick uses a common one for masonry, which is to match the adjacent stonework. The Old Serpentine Works specification gives a typical mortar mix for tendering purposes but alerts the contractor that the final mix will be the result of the approved test panels.

Hardwick Hall, Doe Lea, Chesterfield, Derbyshire

Figure 2.1 Hardwick Hall. © National Trust Images/Andrew Butler.

CONTEXT

Hardwick Hall is a spectacular Grade I-listed Tudor treasure house built in 1590 for Elizabeth, Countess of Shrewsbury, better known as 'Bess of Hardwick' and Elizabethan England's second most powerful and wealthy woman. The building's towers bear her silhouetted initials 'E S'. Hardwick Hall is perhaps the perfect Elizabethan 'prodigy house' – a collection of marvellous buildings created during a construction boom fuelled by the new wealth derived from the former monastic estates. To design it, Bess chose Robert Smythson, a mason by training and one of the first Englishmen to be described as 'architect', and she poured a lifetime of her own experience of building into the project. The building is regarded as one of the acmes of Elizabethan 'renaissance' architecture in England.

Although the exterior is relatively plain, what makes Hardwick Hall unforgettable are its height and symmetry, the ever-changing silhouette of its six towers, and the huge expanses of window glass, inspiring the famous rhyme 'Hardwick Hall – more glass than wall'. The hall was deliberately designed to symbolise wealth and power. Its proportions and the materials used in its construction were intended to support this purpose, and nowhere is this more apparent than in its use and treatment of stone. The scale of the stone is immense: for instance, the spine wall that supports the main staircase and the walls of the six turrets are 1.4 m thick.

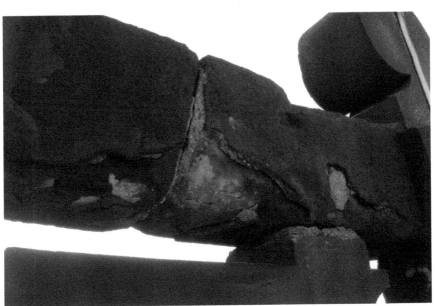

Figure 2.2 (a,b) Recent stone repair work seen alongside original Elizabethan stonework shows clearly the effects of centuries of weathering and modern pollution. © Rodney Melville and Partners.

Bess of Hardwick's wealth enabled her to face the whole of the building in square-hewn stone (ashlar). In the 16th century, stone was the material of choice and Bess was particularly fortunate in being able to quarry high-quality sandstone on the Hardwick Estate. Once broken from the quarry face, the large blocks were reduced in size and then

Figure 2.3 As with the original Elizabethan stone, replacement stone is quarried on the Hardwick estate. © Rodney Melville and Partners.

Figure 2.4 One of the Trust's Direct Labour team of stonemasons. © Rodney Melville and Partners.

carried by packhorse up the hill to the house. Here they were shaped by skilled masons, one of whose first jobs would be to smooth the stones' rough surfaces. Each of the carefully crafted blocks was then levelled on a bed of mortar containing slivers of oyster shells. The lime mortar was also manufactured on the estate, being burnt in kilns in the north orchard, now the car park.

Figure 2.5 The silhouetted initials of Bess of Hardwick, Countess of Shrewsbury, restored to their original glory. © Rodney Melville and Partners.

When Bess died in 1608, the Hardwick Estate passed into the Cavendish family. They chose Chatsworth as their principal home and Hardwick Hall remained a 16th-century showpiece. In 1811, the bachelor 6th Duke of Devonshire inherited the hall. Much of what now exists is the result of his efforts to enhance and preserve the Elizabethan romance of Bess's house. The Duke began the campaign of textile conservation which was taken up in the 20th century by the last member of the family to live there. The Trust acquired the Hardwick Estate from the 11th Duke through the National Land Fund in 1959. This requires it to keep the property in perpetuity and to open it to the public.

The hall was built at the limits of Elizabethan structural engineering and did without the lateral prop of flying buttresses that were used to support the glass-filled walls of earlier buildings, such as medieval cathedrals. As a result, the walls of the rooftop turrets have started to spread. The building has also been damaged by pollution; until the 1980s, Hardwick Hall was surrounded by coal mines and coke works. The pollution that these generated, driven by scouring rain, eroded the hall's stonework, which was not helped by unsuccessful attempts at repair in the early 1900s.

In 1965 the Trust began a new programme of repairs, taking sandstone from the same quarry that Bess of Hardwick had used. The repair programme, funded by the Heritage Lottery Fund, Historic England and the Trust, was completed in 2005, but there is still much work being done to outbuildings.

With help from the Heritage Lottery Fund, the Trust has invested substantial time and money at Hardwick Hall to improve the visitor experience, establish the new Stone and Park Visitor Centres and maintain the fabric of the buildings. This investment has led to important economic and skills benefits.

CONSERVATION PRINCIPLES APPLIED

Owing to the location of the mansion and friable nature of the stone, the lifespan of the masonry is very limited, particularly on exposed elevations. This affects the conservation principles involved in its repair; for example, refacing is not viable as it leads to prema- ture failure so complete stones are often the only option. Like Bess, the Trust has its own in-house dedicated team of stone masons at Hardwick, led by a master mason. The team has the skills necessary to make structural decisions and access to the original quarry.

The masons try to preserve as much as they can and if possible, 'indent' or cut out damaged stone and fix a new piece with lime mortar. In mixing the mortar, the masons need to know about the different materials, colours and textures needed for each type and position of the stone. It is becoming possible to closely date areas of work from the corpus of mason's marks or even the type of stone used.

The approach and specifications are for work on Hardwick Hall itself and the stable- yard cottages. The approach to the project has emphasised one of the most important and generally accepted principles for satisfactory repairs to historic buildings: that newly introduced materials need to be compatible with the old. Lime mortar, render and plaster repair specifications for old buildings therefore should take into account the variable nature of the materials originally used, the effect of weathering, decay over time and the effect on the adjacent building fabric. These considerations also need to allow for environmental and building exposure. A fine balance must be drawn between achieving the strength required for a mortar to withstand varying weather conditions and for the same mortar to be fully compatible with the sometimes delicate, and possibly eroding, masonry that is to be conserved.

The approach at Hardwick Hall was one of the first to assess the principal areas where different mortar types were required; for example:

- bedding and pointing mortar for plane walling, mostly ashlar
- pointing mortar for cills, parapets, copings and other weathering stones
- work at ground level
- masonry mortar repair mixes for ashlar faces
- masonry mortar repair mixes for carved and modelled work
- fillets to fixed window glass lines.

Following the initial survey and assessment of the existing masonry and its defects, the specification process was undertaken in three stages.

1. The general specification for tender purposes
2. The particular specification following full scaffold access
3. The detailed specification following the research and laboratory testing of samples

Included in the Hardwick case study are annotated photographs that help refine the location and details of the specification. Annotated drawings have previously been used to clarify specifications in a visual way and these photographic representations of the repair work provide the contractor with information that is even more precise.

REPAIR SPECIFICATION FOR MASONRY REPAIRS AT HARDWICK HALL

Part A: Masonry – repair work

1. General requirements

1.1 This specification is to be read in conjunction with (and forms an integral part of) the contract documents.

1.2 Where applicable the contractor shall comply with the Construction (Design and Management) Regulations 1994.

1.3 The contractor shall comply with all legal obligations currently in force, and in particular with the Manual Handling Operations Regulations 1992.

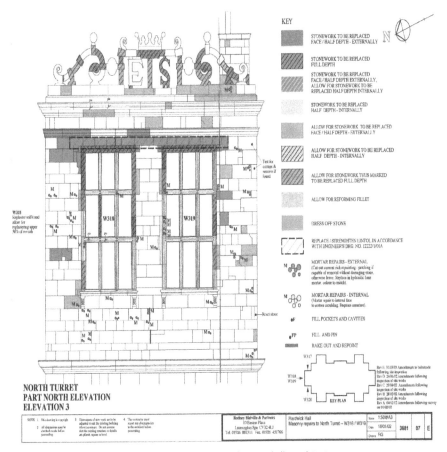

Figure 2.6 Hardwick Hall specification diagram. © Rodney Melville and Partners.

2. Standards

2.1 Generally comply with the following standards: BS 5642, Parts 1 and 2 specification for cills and copings; BS 5628, Part 1 structural use of unreinforced masonry; BS 5390, Stone masonry, subject to any qualifications given below.

2.2 Samples to be provided as specified for approval in accordance with the contract documents.

2.3 Working drawings to be provided as specified in accordance with the contract documents.

3. Materials for repair work

3.1 Stone

3.1.1 Stone is to be from the Hardwick quarry and is to match the approved samples and to be in accordance with the following clauses.

3.1.2 Free from vents, cracks, fissures, discolouration and other defects that adversely affect strength or appearance. **Stones exhibiting strong iron veining are not to be used for repairs to window masonry.**

3.1.3 Free from any defects on the faces visible in the completed work that could mar the appearance of the stonework.

3.1.4 to 3. 17 Standard specification clauses omitted.

3.1.8 Stone from only one quarry bed is to be used throughout.

3.1.9 Stone is to be seasoned at the time of delivery to suit the purpose for which it is to be used.

3.1.10 Stone for ashlar work to be worked to suit the location where it is to be fixed. Finish to match the existing adjacent stonework as approved sample on all faces visible in the finished work.

3.1.11 to 3.1.15 Standard specification clauses omitted.

3.1.16 Stone for carving work to be specially selected to be consistent in grain and colour throughout. Supply sufficient quantity of stone for the carver to have ample scope for his/her work.

3.1.17 Samples of the specified stones are to be submitted for matching purposes. The samples are to match the existing stone in colour, texture, shape, size and surface finish as closely as possible. Where a particular stone has not been specified, the stone samples should be closely related geologically to the stone to be matched.

3.1.18 Standard specification clauses omitted.

3.2 Mortar

Refer to separate specification entitled 'Mortar and pointing'.

3.3 Metal fixings

3.3.1 Dowels, cramps and the like to be non-ferrous grade 304 Austenitic stainless steel.

3.3.2 to 3.4 Standard specification clauses omitted.

3.5 Transport, storage and handling of materials

3.5.1 Transport stone with minimum handling. Stack carefully in vehicle with non-staining packing material to prevent damage.

3.5.2 Store stone on a clean, dry, free-draining surface and prevent contact with soil.

3.5.3 Fairfaced stone is to be stored in stacks on battens, protected with non-staining packing material. Protect from moisture penetration and freezing with non-staining material. Stack stone in such a way as to minimise handling as pieces are selected and transferred for building in on site.

3.5.4 Provide adequate lifting plant to unload and handle stones into position.

3.5.5 Handle stones with tackle, crane or other mechanical aids as necessary.

4. Workmanship for repair work

4.1 Setting out and working drawings

4.1.1 Setting out points are to be taken from the existing building prior to carrying out repair work. Survey reference points should be taken and recorded to enable the correct placement of repairs.

4.1.2 The architect is to be advised of any discrepancies in the alignment of existing work prior to carrying out repairs.

4.1.3 Standard specification clause omitted.

4.1.4 Working drawings, where specified, are to be prepared and submitted to the architect at least two weeks before cutting stonework.

4.2 Preparation of stone

4.2.1 Unless specified otherwise, prepare stone so that the natural bed is:
- horizontal in plain walling
- vertical and at right angles to the wall face in cornices and other projecting stones
- horizontal in quoin stone or end stones which are to be specially selected compact non-laminated stones
- parallel to the radial centre line of each stone and at right angles to the wall face in arches.

4.2.2 Mark the natural bed clearly on each stone before it leaves the quarry. Where it is not possible to determine the bed of the stone after extraction, the bed should be marked before quarrying and subsequently marked on each stone as cut.

4.2.3 Stones are to be cut to full dimensions to match the existing joint layout and width of the existing masonry. Sufficient stone is to be allowed for, to enable working to existing bed and perpend joints, in particular on quoin stones.

4.2.4 to 4.2.5 Standard specification clauses omitted.

4.2.6 Mark each stone on an unexposed face, clearly, to indicate its position in the finished work.

4.3 Working stone

4.3.1 Generally where new stone is being installed to replace existing stonework, profiles and finishes are to match the existing. Where existing stones are weathered, profiles are to be agreed with the architect before proceeding. Do not assume that existing stone profiles will be the same around the building. Stones are to be cut and profiled as necessary to suit their particular location.

4.3.2 Work the exposed and joint faces of each stone to suit its location. Exposed and joint faces to be free from hollow or rough edges. Any saw marks are to be removed from exposed faces; where necessary, this work is to be done prior to construction to ensure accurate alignment of the finished work.

4.3.3 Cut and carve stone to accurate clean profiles before building in. Final dressing off to be carried out *in situ* to ensure accurate alignment with existing adjacent stones. On no account dress off existing stones to suit new.

4.3.4 Chase out mortices for joggles, dowels and cramps before delivery of stone to site.

4.3.5 Drill holes for dowels to be fixed to equal depth in adjacent stones.

4.3.6 Where particularly specified, cut 'V ' shaped sinkings to coincide with those in the ends of adjacent stones for filling with mortar or lead, as specified, to form joggles.

4.3.7 Where particularly specified, cut dovetail channels in adjacent stones for cramps.

4.3.8 Where particularly specified, cut holes for anchor bolts of just sufficient tolerance for easy bolt entry.

4.3.9 New stones are to be worked to dimensions and profiles of existing. The pattern of joints in the existing stonework is to be matched unless specified otherwise.

4.3.10 to 4.3.14 Standard specification clauses omitted.

4.3.15 No work involving preparation or laying of mortar shall be carried out during adverse weather conditions of below 5°C.

4.4 Laying and jointing

4.4.1 Lay stones on their natural bed, or as described in clauses for 'Preparation of the Stone'.

4.4.2 Lay stones on a full even bed of mortar, as agreed sample. Ensure that no hard lumps or the like are present in the mortar that could prevent even bedding.

4.4.3 Horizontal and vertical joints are to be of widths to match the existing masonry, finished as approved sample, unless specified otherwise.

4.4.4 Standard specification clause omitted.

4.4.5 Ensure stone faces taking mortar are dampened as necessary to prevent premature drying out of the mortar.

4.4.6 All joints and joggles are to be filled solid.

4.4.7 All dowels, cramps, ties, metal fixings and the like are to be solidly and securely fixed as the work proceeds.

4.4.8 to 4.4.9 Standard specification clauses omitted.

4.5 Cutting out decayed stonework

4.5.1 Where stone replacement is specified, cut out decayed stonework to accept new stones of minimum depth of 100 mm. Note: stone depth is a 'minimum' requirement. Where necessary, stones are to be of sufficient depth to ensure that the existing pattern of joints is preserved and that carved or moulded stones are adequately bedded. In particular, quoins, jambs, copings and the like are to match existing sizes.

4.5.2 Where 'piecing in' is specified, cut out decayed stonework to specified extent to accept new stone of minimum depth 75 mm and otherwise as necessary to suit the location. Where piecing in is specified, joint widths are to be kept extremely fine.

4.5.3 The contractor is to notify the architect once the whole of the area to be repaired is fully accessible. The architect shall indicate on site any variations to the work found to be necessary.

4.5.4 Where dressing back is required, the contractor shall inform the architect if, in his/her opinion, it is necessary to dress back further, i.e. to a greater depth, than specified.

4.5.5 Stonework is to be adequately propped during the course of the works.

4.5.6 All iron cramps and the like encountered during the course of the works are to be reported to the architect who will decide whether they are to be removed and replaced with non-ferrous fixings or retained, treated and protected.

4.5.7 Stonework is to be cut out with extreme care to ensure that no damage occurs to existing adjacent stones that are to be left *in situ*. In particular, techniques that cause excessive vibration should be avoided.

4.5.8 Where the contractor anticipates that cutting out will not be possible without causing damage, the architect is to be advised before proceeding with the work.

4.6 Standard specifications clause omitted.

4.7 Mortar mix

4.7.1 The mortar should be a little softer and more porous than the stone itself. It should match the original area identified by the architect as closely as possible in colour, texture and strength.

4.7.2 Refer to specification for 'Mortar and pointing'.

4.8 Loose stone to outer face of flues

4.8.1 Where sound single stones are loose, the stone is to be slate-pinned followed by repacking the joints with mortar. If the joints are too fine for repacking then the joints are to be grouted to ensure that all voids around the stone are filled.

4.8.2　Where sound stones are sufficiently loose to make removal comparatively easy (generally where several adjacent stones are loose) and repacking the joints with mortar is not practicable, then grouting should be carried out as above. However, to aid the flow of grout, stones that are removed are to be grooved parallel to the exposed face on all concealed surfaces.

4.8.3　Before any loose stone is removed the position of the stone is to be carefully recorded and the stone rebedded in the same position.

4.8.4　Grouting is to be carried out using hydraulic lime as specified for the adjacent pointing. In cases of wider joints, cut grooves or large voids, silver sand and/or pozzolanic material may be required to be added to the hydraulic lime. The architect should be consulted in each circumstance and an allowance made for trials to ascertain the final mix.

4.9　Mortar repairs to defective masonry

4.9.1　The required mix, colour, finish, extent and method of application of each mortar repair will be specified by the architect upon detailed inspection. The contractor shall notify the architect once the whole of the area to be repaired is fully accessible. The following clauses (4.9.2 to 4.9.9) are to be regarded as a general guide within which variations will be specified.

4.9.2　Cut back decayed stone to minimum depth 25 mm, or as directed by the architect, the slot being dovetailed in section where possible. Saturate with water to reduce suction.

4.9.3　Where instructed by the architect (normally where depth of repair exceeds 50 mm or where repair includes arrises or corners), introduce ceramic armatures in preference to 8 gauge soft copper wire, stainless steel wire or stainless steel Spiro-tie reinforcement, set in pre-drilled holes in sound stones using epoxy adhesive.

4.9.4　The mix to be as directed by the architect on site to suit the conditions. In general, the proportion of binder (lime) to filler (sand/aggregate) is to be in the region of 1:2.

4.9.5　Generally, the following grading of sand and/or stone dust is required to be incorporated in the mix although this will vary in relation to the size of the repair:

Sieve size	% Retained in sieve
1.18 mm	10
600 microns	25
300 microns	25
150 microns	30

Not more than 10% of total volume should pass the 150 micron sieve. Sand and stone dust must be dry before sieving.

4.9.6   The mix is to be applied in coats, built up to the correct profile and finished with a wood float. It must not be over-trowelled which brings laitance on the surface and increases the risk of crazing.

4.9.7   After the initial set, the final texturing to match the repaired stone is applied by a suitable wooden tool.

4.9.8   The finished surface must be protected from the sun for up to 14 days by means of hessian kept damp. It may be necessary to extend this period in exceptionally hot weather.

4.9.9   Where whole stones are to be repaired, each stone is built up individually. Joints between stones are to be pointed at the same time as the work in the mix being employed for general repainting.

5. Protection and cleaning

5.1   Cover arrises, mouldings, carvings and other finished work with adequate protection, securely fixed without damaging new or existing fabric to ensure it is not damaged.

5.2   Keep face work clean and free from staining during construction until completion.

5.3   Clean off and leave stonework clean to the satisfaction of the architect as the scaffolding is taken down.

5.4   All work is to be adequately protected from frost damage. Any frost-damaged work is to be cut out and replaced at the contractor's own expense.

5.5   The work is to be protected from direct sun and rain and against rapid or excessive drying until the mortar is properly cured.

5.6   Turn back scaffold boards adjacent to finished faces to prevent splashing during heavy rain.

5.7   Any mortar or stains caused by the works on the face of the masonry must be completely removed before the mortar hardens.

6. Completion

6.1   The architect is to be advised one week in advance of when the works will be complete in order that an inspection can be made before the removal of any scaffold.

6.2   All dust marks, staining, scuff marks and the like are to be carefully cleaned off. Extreme care is to be taken to ensure that no damage occurs to the new or existing fabric of the building. Any damage is to be made good at the contractor's own expense.

6.3   Remove all tools, plant and equipment and used materials and debris.

6.4   Leave all neat and tidy to the satisfaction of the architect on completion.

Part B: Mortar and pointing

1. General requirements

1.1   This specification is to be read in conjunction with (and forms an integral part of) the contract documents.

2. Standards

2.1 Generally comply with the following standards: BS 5628 Part 3, Code of Practice for use of masonry; BS 5390, Stone masonry, subject to any qualifications given below.

3. Materials

3.1 Sand

3.1.1 The sand shall comply with BS 1200, being clean, sharp and coarse. For repair work the sand shall be of the correct colour and texture so that the new mortar, when dried out, will match the original colour and texture of the original or existing mortar (as appropriate) and the approved mortar sample. The aggregate size is to be well graded to match as closely as possible the existing and to be appropriate to the width of the joint.

3.1.2 The sand shall be free of clay, silt, organic matter and excessive fines. At least half its content shall be a quartz sand. Blending sand and aggregates from different sources may be necessary to achieve the following retentions on a set of standard sieves appropriate for an 8 mm wide mortar joint. All sand and aggregate is to pass a 5 mm sieve except where mortar joints are wider than 15 mm when the aggregate size is to be increased accordingly.
- 10% on a 2.36 mm sieve
- 15% on a 1.18 mm sieve
- 20% on a 600 micron sieve
- 25% on a 300 micron sieve
- 25% on a 150 micron sieve
- 5% only finer than 150 microns

Where narrower or wider joints occur, the grading will be adjusted to suit after discussion with the architect. As a general rule, the largest aggregate particle size should be one-third of the mortar joint width.

3.2 Aggregate

3.2.1 The word 'aggregate' in this context includes one or more of the following: sand, broken and crushed stone and crushed brick. It also includes other crushed inert materials. If sand particles are not of the correct colour or size, or the aggregate in the existing mortar to be matched is composed of crushed or broken stone or other material, this is to be matched with a similar material.

3.3 Water

3.3.1 Water shall be clean and fresh, free from organic and harmful matter in such quantities as would adversely affect the properties of the mortar. Test as directed any water not obtained from the mains.

3.4 Lime

3.4.1 The lime shall be as particularly specified by the architect to suit the location.

3.4.2 **Non-hydraulic lime:** to comply with BS 890: 1995 and BS6463: 1984/1987 and pr EN 459-1, 2 and 3, November 2000.

3.4.3 **Lime putty:** non-hydraulic lime putty may be obtained from an approved supplier or prepared by the traditional method of slaking lump lime. Lime putty must comply with BS 890.1995 and BS 6463: 1987. Hydrated non-hydraulic lime is *not* to be used unless specified.

3.4.4 **Hydraulic lime:** hydraulic lime to comply with the French standard NFP 13-310 and the European standard pr EN 459-1, 2 and 3. The particular class of hydraulic lime, C1, C2 or C3, will be specified by the architect appropriate to conditions of existing masonry, exposure and anticipated weathering. Where specified, hydraulic lime is to be obtained from the following supplier:

Moderately hydraulic lime: Blue lias lime, supplied as bagged dry hydrate from Hydraulic Lias Limes Ltd, Melmoth House, Abbey Close, Sherborne, Dorset DT9 3LH.

3.4.5 The supplier is to provide a current British Board of Agrément certificate for the lime products used. This is available upon request, from the British Board of Agrément, P.O. Box 195, Bucknalls Lane, Garston, Watford, Herts WD2 7NG.

3.4.6 The contractor is to retain a 0.2 mm mesh (200 micron) sieve on site at all times to enable fineness of the dry hydrate to be checked by sampling before use.

3.4.7 The hydraulic lime supplier is to provide a chemical analysis and mortar crushing strength test results, using the dry hydrate currently produced, on request.

3.4.8 **Pozzolanic material:** where specified, is to be ground brick or tile dust, obtainable from Bulmer Brick and Tile Co., Bulmer, Nr. Sudbury, Suffolk CO10 7EF or Metastar 501 obtainable from English China Clays plc, Par Moor Laboratories, c/o John Keay House, St Austell, Cornwall PL25 4DJ. Particle size of brick dust and colour appropriate to specific site conditions to be given following preparation, curing and laboratory testing of samples by the contractor.

3.4.9 **Additives:** no additives (pigments, plasticisers and the like) of any sort shall be incorporated in the mortar except in very special circumstances on express written instructions given by the architect.

3.4.10 to 3.4.11 Standard specification clauses omitted.

3.5 Transport, storage and handling of materials

3.5.1 Approved sands shall be stored in clearly marked bags or bunkers and protected from inclement weather, site debris, leaves and the like. Ensure sand is properly drained before use. Quantities shall be based on the use of dry sand and accurate allowances made for bulking.

3.5.2 Lime putty is to be stored in clean containers, protected from contamination and kept in an excess of water to prevent carbonation.

3.5.3 Hydraulic lime is to be supplied in bags which must be date-stamped with the day of manufacture. Only fresh lime is to be incorporated in the mix from bags opened on the same day as use. Hydraulic lime over 3 months old must not be used. All bags must be delivered to site undamaged and dry without moisture penetration of the covering. Dry hydrate that has been exposed to the air or moisture in transit and in damaged bags is to be rejected and immediately removed from site.

3.5.4 Hydraulic lime must be stored in a dry weatherproof building with a raised floor. Record the date of delivery. No materials may be stored on the ground. Do not store bagged lime on site for more than four weeks.

4. Workmanship

4.1 General

4.1.1 The mortar shall be a little softer and more porous than the stones/bricks themselves and shall comply with BS4551: 1980 for mortars, screeds and plasters and BS4550: 1989 Part 3 sections 3.6 for methods of testing.

4.2 Samples

4.2.1 A range of sample mortar mixes are to be prepared for inspection by the architect. For repair work the mixes are to match the existing mortar in respect of colour, texture, strength and mortar joint width. The architect will identify an area of existing mortar to be matched.

4.2.2 The contractor is to allow for blending and sieving sands from different sources, as necessary, to achieve appropriate grading, colour and texture.

4.2.3 If the architect does not consider the sample mixes to be satisfactory in respect of the requirements of this specification, then further samples are to be provided by the contractor as necessary.

4.2.4 Where applicable and sound existing pointing survives in good condition, the architect will mark a section which is to be matched in respect of colour, texture, strength and mortar joint width.

4.2.5 A trial area of pointing is to be executed on specially prepared sample panels (NOT THE EXISTING BUILDING FABRIC) for approval by the architect using mortar as specified and approved. The precise location for this trial must be agreed with the architect before proceeding.

4.2.6 Where directed, the contractor shall prepare a sample panel of masonry, executed in mortar mixes and pointing techniques as specified by the architect, or to match existing work.

4.3 Mortar types and mix

4.3.1 Mortar mix: only after a sample has been approved by the architect shall the contractor prepare aggregate blended to the correct grading and/or coarse stuff. The contractor should therefore be aware of the subsequent preparation and standing times prescribed in this specification for the various materials.

4.3.2 Mix materials sufficiently to obtain a uniform colour and consistency and as specified elsewhere.

4.3.3 **Lime putty:** lump lime shall be properly slaked in an excess of water and stirred and hoed to ensure coagulation does not occur. The putty shall then be screened through gauze equivalent to a 2.36 mm sieve to BS 410. Non-hydraulic lime shall then be left for at least 48 hours. Prior to batching all excess water shall be siphoned off. It shall then be matured for not less than three months. Where specified, hydrated non-hydraulic lime shall be soaked to putty by mixing with water and allowing to stand for not less than 24 hours, then the excess water drained off for use as above. The putty at time of use to be strictly in accordance with BS890 and BS6463.

4.3.4 **Coarse stuff (non-hydraulic lime):** the coarse stuff of one part non-hydraulic lime putty or quicklime granules to three parts sand shall be batched by gauge box. If it is mixed in a mechanical mixer only the minimum amount of water to achieve mixing shall be added. The minimum time to be taken for mixing each full batch of all ingredients is 20 minutes in a mechanical mixer. Paddle mixers are preferred. Roller pan mixers are not to be used unless the setting of an adjustable roller height from the base of the pan can be demonstrated and agreed with the architect before proceeding.

4.3.5 The coarse stuff of non-hydraulic lime shall then be kept covered with damp sacking or polythene sheeting to prevent drying out, and stored for a minimum period of two weeks.

4.3.6 *Standard specification clause omitted.*

4.3.7 **Hydraulic lime:sand mixes:** mortar is to be prepared using hydraulic lime and well-graded aggregate of the appropriate particle size to suit the joint width. The qualifications for mixing are the same as those set out in paragraph 4.3.4 above.

4.3.8 The following basic mixes should be assumed for tendering purposes subject to confirmation by the architect following preparation of samples and testing. Allow for making minor adjustments to these mixes in order to match the existing mortars for colour, texture and durability.
- 1:2 moderately hydraulic lime: graded aggregate for plain walling.
- 1:2 eminently hydraulic lime or moderately hydraulic lime gauged with pozzolan: graded aggregate for copings, chimneys and other exposed locations including work adjacent to and below ground.

4.3.9 Before work commences on site the architect will provide detailed specifications for mortars to suit the particular conditions of use. In exposed locations where the stone is hard and durable an eminently hydraulic lime or moderately hydraulic lime with pozzolan may be used. Where the stone is very soft and the exposure low, a feebly hydraulic lime or far lime (with or without pozzolanic additive) may be used subject to the work being carried out at an appropriate time of year. The addition of pozzolanic material, if required, will be specified by the architect following receipt of test results. Bedding mortar may vary from pointing material where directed.

4.3.10 All mixing and handling equipment is to be kept clean. Containers, boards, tools, etc. shall be well cleaned before the next batch of mortar is mixed/used.

4.3.11 **Measures**: constituents by volume in clean gauge boxes. Cement gauging boxes are to be kept dry at all times. Gauging by the shovel is strictly forbidden.

4.3.12 **Mixing:** hydraulic lime, as specified, is to be added to moderately wet aggregate when knocking up the appropriate mortar mix. The addition of water must be kept to the minimum. Additional work ability if necessary may be achieved by increasing the mixing time *not* by adding water. A mechanical mixer may be used at this point and the minimum period recommended for continuous mixing is 20 minutes. All mortar is to be used within 24 hours of mixing for blue lias hydraulic lime (HL/2) only. The length of time between mixing and use is to be consistent and agreed with the architect before proceeding.

4.4 Preparation

4.4.1 In preparation for repointing works, using hand tools only, rake out all loose jointing material to a depth of not less than twice the joint width. All raking/cutting shall leave a clean, square face at the back of the joint, so as to provide optimum contact with the new mortar.

4.4.2 The prepared face and joint should be carefully cleaned out with a bristle brush and thoroughly flushed out with clean water, avoiding unnecessary saturation. All dust and loose material must be removed, working from top to bottom of the wall.

4.4.3 All cutting out and cleaning works should be approved prior to commencement of the repainting.

4.4.4 No cleaning agents or fungicides are to be used either before or after repair works, except on the express authority of the architect.

4.4.5 Any sound pointing should be left undisturbed, even if it has weathered back behind the general wall face to as much as half the joint width. Generally, the existing mortar should be capable of being removed by raking out by hand with a blunt instrument, leaving the arrises of the brickwork or stone unharmed. A hammer and chisel should not be used unless permitted by the architect. Under no circumstances should an angle grinder or similar tool be used.

4.4.6 Where it is desirable to remove damaging and unsightly cementitious pointing, experiments should be carried out to the approval of the architect to ascertain the most appropriate method of removal and to limit damage caused to existing work.

4.5 Pointing

4.5.1 It is essential that the masonry is thoroughly dampened when pointing is commenced. If the joints have dried out before cleaning they must be rewetted with a hand-held spray prior to placing of any new mortar. No water should be left lying within the prepared joint.

4.5.2 The mortar should be pushed into the joint and firmly ironed in with the maximum possible pressure and minimum of over-working. Pointing irons should be used, not pointing trowels. The pointing irons may be cranked, bronze or steel flat of a width which will fit into the joint and ensure compaction over the full width. Compaction is therefore achieved throughout the depth of the joint each time mortar is placed rather than from the surface alone. The contractor should be aware that it may be necessary for him to fabricate pointing irons to undertake the works.

4.5.3 If the joint is to be filled in one operation, the mortar must be almost crumbly and be ironed in very firmly.

4.5.4 Repointing work should begin at the uppermost section of the wall and proceed downwards, ensuring that all the mortar is pressed well into the joints to achieve good compaction. Fill all the joints solidly with the approved mortar mix finishing very slightly back from the masonry and in accordance with the approved sample. The mortar is not to be spread or buttered onto the face of the masonry unless specified.

4.5.5 Where directed to produce a weathered, roughened finish to expose the aggregate, the mortar should be left to take its initial set and then stippled with a stiff bristle brush. The bristles should not be dragged across the face but tapped against it. Timing is critical. If this technique is applied too early the mortar will be removed too easily and the bond forming between the brick and the mortar will be disrupted. If too late, it will be difficult to make the required impression. Light damping of the surface to remove laitance from the surface of the aggregate at this point is permissible.

4.5.6 The particular finish required (stippled, struck, double struck, tuck pointed) is to be specified by the architect.

4.5.7 Any slight fractures due to shrinkage must be cut out and remade.

5. Protection and cleaning

5.1 General

5.1.1 The work shall be protected from direct sun and rain and kept well ventilated and moist at an average daily temperature of not less than 10°C for a minimum of four weeks until the face has dried and hardened to ensure satisfactory 'curing'.

5.1.2 Protect all new work against frost; any joint damaged due to frost action is to be cut out and redone in frost-free conditions.

5.1.3 During dry weather all new pointing shall be kept continuously moist for a minimum of four weeks (but not wet) to ensure that the set takes place slowly.

5.1.4 Turn back scaffold boards adjacent to brick faces at night or during heavy rain.

5.1.5 Any mortar or stains caused by the works on the face of the masonry must be completely removed immediately and before the stains dry or mortar hardens.

5.2 Frost

5.2.1 Work must not be carried out during frost conditions. Frost must not be allowed to affect completed work before it has fully cured. Manufacturers of hydraulic lime are to guarantee the curing periods for the class of lime supplied.

5.2.2 Generally pointing and bedding mortar is not to be laid when the temperature is 5°C or below and falling. Work may recommence when the temperature is 3°C or above and rising. The contractor is to keep thermometers on site to record maximum and minimum temperatures for the duration of the contract and to keep a daily record of night and daytime temperatures for winter working as directed.

6. Completion

6.1 The architect is to be advised at least one week in advance of when the work will be complete in order that an inspection can be made prior to the removal of any scaffold.

6.2 Remove all tools, plant and equipment and used materials and debris.

6.3 On completion leave all neat and tidy to the satisfaction of the architect.

Specification author: Rodney Melville and Partners, Architects and Historic Building Consultants

COMMENTS ON THE RESULT

The overall objective at Hardwick Hall was to retain as much of the original fabric as possible, working to the constraints of using 'Hardwick stone' (coal measures sandstone). The detailed drawings and specification were essential for the works to be priced efficiently. These drawings and images also provided an excellent starting point to record the work undertaken. The knowledge of the in-house surveyors and the master mason should not be ignored on seeking their advice and comments on the specification prior to tender issue.

A full appreciation and understanding of the environment to which the masons were working were essential in that walls were not plumb, out of line and not square. So when new openings were reinstated or items of masonry were banked, these considerations needed to be taken into account. One of the challenges facing the team was that the remaining good-quality stone in the Hardwick quarry was severely depleted so a replacement stone had to be sourced.

Work was undertaken to the voussoirs to one of the link archways between the clock tower barn and ox house. The faces of the eroded voussoirs were removed and replaced with new stone, an amazing piece of work.

The masonry project is a great success and credit must go to the quality of the specification and the execution of the work by the masons which proves the point that if you have a quality specification and exceptional trade personnel, the results will speak for themselves.

Thorington Hall, Stoke by Nayland, Suffolk

CONTEXT

Thorington Hall is a rambling Grade II* farmhouse set in the heart of the Stour Valley. It is one of the largest and most beautiful of Suffolk's timber-framed farmhouses. It is a traditional oak-framed building, erected on a brick plinth, with roof coverings of plain tiles and gabled exterior. The house is the product of three distinguishable structural phases: there is evidence of a late 15th- or early 16th-century building but the house as it stands today is largely the product of massive rebuilding in 1620–30, with single-storey 18th-century additions.

The Umfreville family occupied the house for some of its life, but little else is known of its early history. The house was given to the Trust by the distinguished medical scientist and mathematician Lionel Penrose in 1940 shortly after the completion of a major pro-gramme of repairs put in hand by the architect Marshall Sisson, in 1937. This restoration reinstated certain ancient features and removed a number of minor 19th-century additions and modifications.

The finest exterior feature of Thorington Hall is the enormous six-flued brick chimney stack, which is primarily responsible for its listed status. The chimney has six grouped octag-onal shafts with spurred, star-shaped caps and moulded octagonal bases. The chimney serves the fireplace in the central range of the building as well as two chambers in the northern part of the west range. It has its origin in the late medieval period and is on a par with the chimneys of Jacobean buildings such as Blickling Hall, Norfolk, and Hatfield House, Hertfordshire.

Figure 2.7 Thorington Hall chimney – after restoration. © Pick Everard.

CONSERVATION PRINCIPLES APPLIED

The instigation for works to the chimney was a small fire in the chimney stack. As part of the post-fire investigation, the Trust surveyor arranged an internal CCTV survey of the flue in which the fire had occurred. Major erosion of the internal structure was found, giving rise to serious concerns over the stability of this very tall brickwork structure. It was apparent that there was a strong risk that the top half of the stack might collapse under high winds, endangering the tenants and the public (in the adjacent highway), as well as damaging the building itself.

A structural engineer was appointed to provide advice on the matter. The engineer agreed with the initial assessment and recommended the implementation of a temporary structural scaffold and shoring system to stabilise and maintain the existing structure. This was quickly implemented, and allowed further investigations to verify the concerns, whilst protecting the structure, tenants and the public.

Although the signs of structural instability were now known, and were the primary trigger for the project, it was also clear that the full extent of the problem could not be established until work was under way. It was only as the stack was disassembled from the top, brick by brick, that the limits of the instability were discovered. In a modern non-conservation repair, one might expect to predict and specify an exact height at which to limit the repair, giving a safe margin for error. However, this being a conservation repair, the aim had to be 'to carry out as little as possible but as much as is necessary' to reach a stage where the scale of the task could be seen. Such unknowns in project scope obviously caused cost uncertainty, a common feature of conservation-based projects and their specifications.

A number of repair options were considered. The fundamental issue was that access to the internal faces of the stack to undertake any effective repair was not physically possible. The very slender nature of the detailing applied meant the normal weathering process of the mortar had compromised the structural stability. Also, previous repairs had been undertaken in non-vernacular bricks. After detailed discussions with Historic England and the Conservation Officer, it was decided that complete reconstruction of the top half of the stack was the only practical option to provide a long-term solution. This necessitated a complete rebuild with only a few salvaged bricks able to be reused on the inner faces. The new bricks were carefully sourced to match:

- size
- colour
- texture
- special shape
- firing technique
- clay composition and source.

The specification emphasises the requirement to record the structure photographically before work is begun and continuously throughout the project. A record of individual masonry units is also required, as is to have them referenced to an overall drawn or

Figure 2.8 View of brickwork corbelling at top of chimney. © Pick Everard.

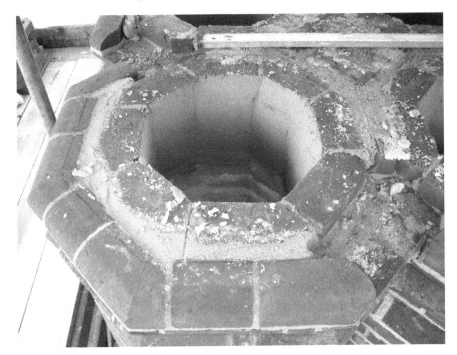

Figure 2.9 View of brickwork corbelling before splitting into six stacks. © Pick Everard.

Figure 2.10 Finished restoration work. © Pick Everard.

photographic plan. This recording specified before, during and after the project helps add to the understanding of the particular area being worked on and the larger body of knowledge about the historic structure in general.

The specification also forewarns the contractor that if any artefacts or remains are discovered in the course of the work, the Trust may instigate an archaeological investigation. This touches on a general conservation principle that should be specified since the type of contractor used, their experience and attitude may be unknown ahead of time. An encompassing legal point should be made within contract documents that all items of archaeological value found in the building during the course of the works belong to the employer. However, the important second conservation point to be specified is that upon the discovery of these items, the contractor should notify the contract administrator and await further instructions before disturbing them.

REPAIR SPECIFICATION FOR CHIMNEY STACK AT THORINGTON HALL

Notes

1. Numbering within this specification relates to the National Building Specification.

2. Specification is required to be read with preliminaries/general conditions.

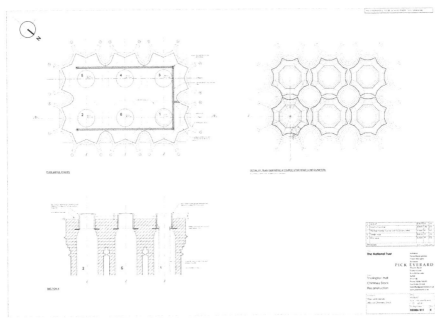

Figure 2.11 Chimney detail drawing. © Pick Everard.

C41 Repairing/renovating/conserving masonry

GENERALLY/PREPARATION

10 INFORMATION TO BE PROVIDED WITH TENDER

- Submit: proposals for the protection of internal finishes, staircase and flooring for the whole of the internal access route/routes within the building.

110 SCOPE OF WORK

- The proposal is to carefully take down the complete stack visible above the roof line extending approximately 1800 mm into the roof void. A daily digital photographic record will be made as the stack is taken down.

The stack is to be taken down brick by brick, carefully cleaning the bricks and setting aside all those reusable. The existing bricks have been laid in a cement mortar and some are very friable. Every effort is to be made to reuse as many existing bricks as possible with the proviso that only sound bricks suitable for incorporation in the works are to be used. Where this is not possible new purpose handmade bricks obtained from the Bulmer Brick and Tile Co Ltd will be used. All remaining bricks to be removed from the site.

The chimney stacks are to be rebuilt using materials appropriate for the age of the building. This includes the use of NHL mortar.

A 200 mm multi-fuel liner is to be installed in one flue to serve a wood-burning stove to be installed in the ground floor room.

The taking down of the stacks will be closely monitored particularly during the closing stages when the condition and structural stability of the brickwork at the lower level will be assessed. It is at this stage that a decision will be made concerning the need to take down and rebuild the stack below the roof level.

The inner faces of the flues are to be parged and low-profile chimney pots set at the top of each flue. Sealed pots are to be used for the redundant flues and the top of the stack will be flaunched in lime mortar to protect the brickwork.

It should be noted that the National Trust will instigate an archaeological investigation should any important remains be uncovered.

It should also be noted that the works will require the careful stripping of the adjacent plain tile roof surfaces including the curved tile valleys and ridge tiles and battens, etc., to allow the dismantling and rebuilding of the stack and the reinstatement of the roof covering on completion.

120 REVIEWING SCOPE OF THE WORK
- Inspection: arrange before starting work. Confirm type and extent of work required.
- Marking: mark clearly, but not indelibly, on face of masonry units or parts of units to be cut out and replaced.
- Identification of masonry units to be removed, replaced or repaired: code number cross-referenced to drawings/photographs.
- Records of masonry to be repaired: before starting work, use measurements and photographs as appropriate to record bonding patterns, joint widths, special features, etc.

WORKMANSHIP GENERALLY

150 POWER TOOLS FOR REMOVAL OF MORTAR
- Usage: not permitted.

160A PROTECTION
- Storage of masonry units: on level bearers clear of the ground, separated with resilient spacers. Protect from adverse weather and keep dry. Prevent soiling, chipping and contamination by salts and other deleterious substances.
- Protection of masonry: suitable non-staining slats, boards, etc. Remove at completion.
  – Prevent damage, particularly to arrises, projecting features and delicate, friable surfaces.
  – Prevent mortar/grout splashes and other staining and marking.
- Protect surfaces of surrounding roof coverings to prevent staining and marking by mortar, etc. for the duration of the works.

165A STRUCTURAL STABILITY
- General: maintain stability of masonry. Report defects, including signs of movement, that are exposed or become apparent during the removal of masonry.

170A DISTURBANCE TO RETAINED MASONRY
- Retained masonry in the vicinity of repair works: disturb as little as possible.

180 OPERATIVES
- General: skilled and experienced with the materials and procedures required.
  - Evidence of training and previous experience: provide on request.

185A ADVERSE WEATHER
- Frozen materials: do not use. Do not lay masonry on frozen surfaces.
- Air temperature: do not bed masonry or repoint:
  - in hydraulic lime:sand mortars when at or below 5°C and falling or unless it is at least 3°C and rising
  - in non-hydraulic lime:sand mortars in cold weather without approval.
- Temperature of the work: maintain above freezing until mortar has fully set.
- Rain and snow: protect masonry by covering during precipitation and at all times when work is not proceeding.
- Hot conditions and drying winds: prevent masonry from drying out too rapidly.
- New mortar damaged by frost: rake out and replace.

190 CONTROL SAMPLES
- General: obtain approval of the following before proceeding with the remainder: sample panels of brickwork (nom. 1 m square) using selected mortars built in locations as directed.
- Protection: protect from adverse weather and damage.

195A SAND SAMPLES
- Approval: before placing order, submit for approval representative samples of sands for bedding, pointing and jointing as Z21/320.

MATERIAL/PRODUCTION/ACCESSORIES

260A BRICKS
- Types: handmade red clay 2 inch facing bricks as supplied and delivered by the Bulmer Brick & Tile Co Ltd.
- Sizes/special shapes: purpose-made bricks: to the profiles to match existing and as shown on the drawings as supplied and delivered by the Bulmer Brick and Tile Co Ltd.

265A SALVAGED BRICKS
- Condition:
  - Free from matter such as mortar, plaster, paint, bituminous materials and organic growths.
  - Sound, clean and reasonably free from cracks and chipped arrises.

270A TERRACOTTA
- Manufacturer/supplier: approved.

- Pattern: low-profile beaded flue terminals to active flues, low-profile beaded flue terminals with sealed tops to non-active flues perforated for ventilation.
- Finish: smooth.
- Colour/texture: to match approved samples.

275A PRODUCTION OF TERRACOTTA

- Produced accurately within agreed manufacturing tolerances.
- Quality: free from imperfections that will have an adverse effect on durability.
- Appearance: to match the range of approved samples.

DISMANTLING/REBUILDING

310A DISMANTLING MASONRY FOR REUSE

- Masonry to be reused: remove carefully.
- Old mortar, dirt and organic growths: clean off and leave masonry in a clean sound condition for reuse.

320A REBUILDING BRICKWORK

- Replacement materials: brick as clause 260A.
- Mortar: as section Z21.
  - Mix: 1:2.5 NHL3.5 hydraulic lime:sand.
  - Sand source/type: BS882 Table 4 C or M. Samples required for approval as Z21 clause 320.
- Rebuilding: to match previous face and joint lines, joint widths and bonding.
- Laying: on a full bed of mortar, and all joints filled.
- Exposed faces: keep clear of mortar and grout.
- Joints: flush.

ANCILLARY ITEMS

910 FLUE LINER TO CHIMNEY FLUE

- From HETAS-approved supplier.
- Product reference: 200 mm multi-fuel flue liner suitable for wood-burning stoves.
- Other requirements: plate and clamp at top of chimney to support flue liner built into brickwork as shown on the drawings and connection to register plate at fire opening.

F10 Brick/block walling

To be read with Preliminaries/General conditions.

TYPES OF WALLING

110 CLAY FACING BRICKWORK TO CHIMNEY STACK

- Bricks: handmade red clay 2 inch facing bricks.
  - Manufacturer: the Bulmer Brick and Tile Co Ltd, Bulmer, Near Sudbury, Suffolk CO10 7EF. Product reference: handmade red clay facings.
  - Special shapes: as shown on drawings.

- Mortar: as section Z21.
  - Standard: not applicable.
  - Mix: 1:2.5 NHL3.5 hydraulic lime: sharp well-graded sand.
  - Additional requirements: submit samples of sand for approval.
- Bond: to match existing.
- Joints: flush.
- Features: as shown on the drawings.

230 RECLAIMED BRICK FACING BRICKWORK TO CHIMNEY STACKS

- Reclaimed bricks: existing bricks salvaged during the dismantling of the existing chimney stack.
  - Condition: sound, free from mortar and deleterious matter.
  - Supplier/source: existing chimney stack.
  - Format: as existing.
- Mortar: as section Z21.
  - Standard: not applicable.
  - Mix: 1:2.5 NHL3.5 hydraulic lime: sharp well graded sand.
  - Additional requirements: submit samples of sand for approval.
- Bond: to match existing.
- Joints: flush.
- Features: as shown on the drawings.

WORKMANSHIP GENERALLY

500 LAYING GENERALLY

- Mortar joints: fill vertical joints. Lay bricks, solid and cellular blocks on a full bed.
- Bond where not specified: half lap stretcher.
- Vertical joints in facework: even widths. Plumb at every fifth cross-joint.

535 HEIGHT OF LIFTS IN WALLING USING CEMENT GAUGED OR HYDRAULIC LIME MORTAR

- Quoins and advance work: rack back.
- Lift height (maximum): 1.2 m above any other part of work at any time.
- Daily lift height (maximum): 1.5 m for any one leaf.

561A COURSING BRICKWORK

- Gauge: as existing brick courses.

580 LAYING FROGGED BRICKS

- Single frogged bricks: frog uppermost.
- Double frogged bricks: larger frog uppermost.
- Frog cavity: fill with mortar.

635A JOINTING

- Profile: flush and consistent in appearance.

645 ACCESSIBLE JOINTS NOT EXPOSED TO VIEW

- Jointing: struck flush as work proceeds.

690 ADVERSE WEATHER

- General: do not use frozen materials or lay on frozen surfaces.
- Air temperature requirements: do not lay bricks/blocks:
  - In hydraulic lime:sand mortars when at or below 5°C and falling or below 3°C and rising.
  - In thin joint mortar glue when outside the limits set by the mortar manufacturer.
- Temperature of walling during curing: above freezing until hardened.
- Newly erected walling: protect at all times from:
  - Rain and snow.
  - Drying out too rapidly in hot conditions and in drying winds.

ADDITIONAL REQUIREMENTS FOR FACEWORK

710 THE TERM FACEWORK

- Definition: applicable in this specification to all brick/block walling finished fair.
  - Painted facework: the only requirement to be waived is that relating to colour.

740 FINISHED MASONRY WORK REFERENCE PANELS

- General: before proceeding to construct the following walling types, construct panels as specified. Give notice when panels are dry.
- Selection masonry units: reasonably representative of the average quality of the whole order to be delivered.
- Panel types:
  - Walling type: F10/110.
  - Location: to be advised.
  - Size: 900 × 900.
  - Other requirements: one course dog-tooth profile special bricks.

750 COLOUR CONSISTENCY OF MASONRY UNITS

- Colour range: submit proposals of methods taken to ensure that units are of consistent and even appearance within deliveries.
- Conformity: check each delivery for consistency of appearance with previous deliveries and with approved reference panels; do not use if variation is excessive.
- Finished work: free from patches, horizontal stripes and racking back marks.

760 APPEARANCE

- Brick/block selection: do not use units with damaged faces or arrises.
- Cut masonry units: where cut faces or edges are exposed cut with table masonry saw.

- Quality control: lay masonry units to match relevant reference panels.
  - Setting out: to produce satisfactory junctions and joints with built-in elements and components.
  - Coursing: evenly spaced using gauge rods.
- Lifts: complete in one operation.
- Methods of protecting facework: submit proposals.

800   TOOTHED BOND

- New and existing facework in same plane: bond together at every course to achieve continuity.

830   CLEANLINESS

- Facework: keep clean.
- Mortar on facework: allow to dry before removing with stiff bristled brush.
- Removal of marks and stains: rubbing not permitted.

M20   Plastered/rendered/roughcast coatings

To be read with Preliminaries/General conditions.

TYPES OF COATING

310A   LIME/SAND PARGING TO FLUES

- Substrate: brickwork.
  - Preparation: not required.
- Lime manufacturer: contractor's choice.
  - Product reference/type: hydraulic NHL3.5.
- One coat:
  - Mix: 1:2–3 with hair reinforcement.
  - Sand: as clause Z21/320A for brickwork.
  - Thickness: 13 mm.
  - Finish: smooth.
- Other requirements: parging applied in course of construction and cored with a sack stuffed with shavings and with a rope attached. The core placed in the flue at commencement and drawn up in stages as the work proceeds to keep the flue clear of mortar droppings, brick debris, etc.

478   HYDRAULIC LIME

- Standard: to BS EN 459-1.
  - Type: natural hydraulic lime (NHL).

492   HAIR REINFORCEMENT

- Manufacturer/supplier: contractor's choice.
  - Product reference: supplier's reference.
- Proportions (approximate): 5 kg hair to 1 m³ of coarse stuff.
- Condition: clean, free from grease and other impurities. Well teased before adding to the mix.

- Distribution: evenly throughout with no balling into lumps.
- Storage period for haired mortar (maximum): four weeks.

495 MIXING
- Render mortars (site-made):
  - Batching: by volume. Use clean and accurate gauge boxes or buckets.
  - Mix proportions: based on damp sand. Adjust for dry sand.
  - Lime:sand: mix thoroughly. Allow to stand, without drying out, for at least 16 hours before using.
- Mixes: of uniform consistency and free from lumps. Do not retemper or reconstitute mixes.
- Contamination: prevent intermixing with other materials.

497 COLD WEATHER
- General: do not use frozen materials or apply coatings on frozen or frost-bound substrates.
- External work: avoid when air temperature is at or below 5°C and falling or below 3°C and rising. Maintain temperature of work above freezing until coatings have fully hardened.
- Internal work: take precautions to enable internal coating work to proceed without damage when air temperature is below 3°C.

Z21 Mortars

To be read with Preliminaries/General conditions.

LIME:SAND MORTARS
320A SAND FOR LIME/SAND MASONRY MORTARS
- Type: sharp, well graded to BS 882 Table 4 C or M for brickwork.

Table 4 Sand

Sieve size	Percentage by mass passing BS sieve			
	Overall limits	Additional limits for grading		
		C	M	F
10 mm	100	–	–	–
5 mm	89 to 100	–	–	–
2.36 mm	60 to 100	60 to 100	65 to 100	80 to 100
1.18 mm	30 to 100	30 to 90	45 to 100	70 to 100
600 um	15 to 100	15 to 54	25 to 80	55 to 100
300 um	5 to 70	5 to 40	5 to 48	5 to 70
150 um	0 to 15	–	–	–

NOTE: individual sands may comply with the requirements of more than one grading. Alternatively, some sands may satisfy the overall limits but may fall within any one of the additional limits C, M or F. In this case, and where sands do not comply with Table 4, an agreed grading envelope may also be used provided that the contractor can satisfy the contract administrator that such materials can produce mortar of the required quality.

- Sand for facework mortar to be washed, coarse pit sand, entirely free from organic matter and to match aggregates in the existing lime mortar, both in material and grain size.
- Provide a 400 gm sample of each aggregate proposed and forward to the contract administrator four weeks in advance of commencement of works on site. For each sand sample, state the region and quarry of origin and the grading and mineralogy of the aggregate. Prepare sample biscuits of mortar of each of the types listed above produced with each of the sand samples offered for approval.
- Sand should be kept covered to prevent washing out of fines, separation and to stabilise the water content.

335A HYDRAULIC LIME FOR MORTAR (NHL)
- Hydraulic lime will be pure natural hydraulic lime to DD ENV Part 1 NHL designation obtained from an approved source.
- Ensure materials supplied have manufacture date on packaging and are not more than three months old.

335B READY PREPARED LIME PUTTY
- Manufacturer: to contractor's choice as approved by the contract administrator.
  – Product reference: mature lime putty.
- Maturation period before use (minimum): 12 months.
- Lime putty shall be kept in its original airtight container with a water film over until needed.

350 STORAGE OF LIME:SAND MORTAR MATERIALS
- Sands and aggregates: keep different types/grades in separate stockpiles on hard, clean, free-draining bases.
- Ready-prepared non-hydraulic lime putty: prevent drying out and protect from frost.
- Non-hydraulic lime:sand mortar: store on clean bases or in clean containers that allow free drainage. Prevent drying out or wetting and protect from frost.
- Bagged hydrated hydraulic lime: hydraulic lime shall be kept in its waterproof wrapper off the ground in dry conditions. Once opened, it must be kept in a dry store and used within a working day. Thereafter it should be discarded.

360A MAKING LIME:SAND MORTARS GENERALLY
- Measure materials accurately by volume using clean gauge boxes.
- Proportions of mixes are for dry sand, allow for bulking if sand is damp.
- Lime putty and sand shall be mixed together thoroughly on a boarded platform, without the addition of water, and the mortar must be chopped and pounded with the edge and flat of the spade to expel excess water and to compress the mix, until the lime is uniformly distributed throughout the mass. Roller pan or other mechanical batch mixers may be used with prior written approval.

- Mix ingredients thoroughly to a consistency suitable for the work and free from lumps.
- Keep plant and banker boards clean at all times.

390   KNOCKING UP NON-HYDRAULIC LIMES AND MORTARS

- Knocking up before and during use: achieve and maintain a workable consistency by compressing, beating and chopping. Do not add water.
  - Equipment: roller pan mixer or submit proposals.

400A   MAKING HYDRAULIC LIME:SAND MORTARS (NHL)

- Mixing hydrated hydraulic lime:sand: follow the lime manufacturer's recommendations for each stage of the mix. In mixing hydraulic lime mortars a whisk attachment to an electric drill has been found useful. Mix for 10 minutes thus.
- Water quantity: only sufficient to produce a workable mix.
- Working time: use mortar within about two hours of mixing at normal temperatures. Do not use after the initial set has taken place and do not retemper.

410   MAKING GAUGED MORTARS

- Ingredients for coarse stuff shall be measured in approved gauge boxes and shall be mixed at least seven days before use and if possible at the commencement of the works, following approval of sands, mortar biscuits and sample panels.
- Mix as clause 360A. Coarse stuff so prepared must be covered and kept damp until required, preferably in polythene sacks with tie heads.
- No gauged mortars will be knocked up for reuse.
- Thoroughly mix the mortar once again before use again without the addition of water unless the mortar appears crumbly, when a small quantity of water may be added to improve plasticity.
- Avoid excess water.

500   LIME MORTARS AND WINTER WORKING

- During the period mid-October to April refer to the contract administrator for confirmation of mix type if high calcium mixes are specified.
- Hydraulic lime mixes require greater care in winter than in summer. To achieve a satisfactory set, 96 hours of constant temperatures above 8–10°C are required. For mortar depths greater than 15 mm this period is extended.
- Application should not be undertaken when temperatures are below 8°C or are rising from this temperature unless a means of creating constant warm air behind a sheeted protection is provided.

Precautions

- Ensure bulk materials and mixing water temperatures are above 8°C at the time of mixing.
- Provide three layers of protection whenever frost is likely. The following should be considered minimum recommendations.

1. Hessian sacking: 12 oz bleached, starch-free hessian in contact with the working surface.
2. Insulation: either loose hessian sacks of hay or straw in a 6 inch layer or minimum four layers of plastic bubble wrap laid or wrapped so that it may be removed when the temperature rises to 8°C to permit necessary air movement to the mortar. Adjust the number of layers according to severity of frost anticipated.
3. Polythene sheeting or tarpaulin to shed water clear of the protective underlayers.

- Adequate protection from frost including the provision of warm air heating remains the responsibility of the contractor.
- Protect during frosty weather for a minimum of five days, removing layers to allow air movement when temperature is above 8°C and reinstating as the temperature drops.
- Frost attack of hydraulic lime mortars will produce a powder with little or no strength. This should be removed for the depth of the damage and replaced with a fresh application which will most likely result in the taking down and rebuilding of the areas affected.

Specification author: Pick Everard

COMMENTS ON THE RESULT

The primary concern for the Trust surveyor at Thorington Hall was the extent of the rebuilding when it was found that, owing to mortar and parging losses, the bricks freely lifted out. There was no adhesion remaining between the bricks and mortar. As such, a significant extent of chimney was only held in place by the self-weight of the structure. The other difficult decision was at what level to stop dismantling and commence reconstruction. At the upper levels of the chimney, the brickwork was half to one leaf thick, so the erosion (up to 100/125 mm) was very significant, together with the loss of adhesion. Once within the roof void, at second-storey ceiling level, the erosion was less but also within a much thicker masonry structure. In addition, at this level, the lime mortar had not been compromised by the weather and still had good adhesion.

Fundamental to the project was having a structural engineer who was confident and experienced with the nature of this structure in order to make the necessary decision to reach the dismantling phase. The rebuilding phase was overseen by an architect, who, prior to the works, had to undertake a significant degree of detailed recording of the existing structure to provide a record for the rebuilding.

The main issue during the construction phase was the supply of the replacement hand-made bricks. The procurement time from the specialist supplier was very tight, and there were a significant number of special shapes within the structure. On a number of occasions, supplied materials were of unacceptable quality or insufficient number. This was generally due to the nature of the product. This did lead to some downtime of the works.

The main lesson learned from the works was that from external inspection at ground level (even through binoculars), this chimney structure had appeared to be in a reasonably fair condition. It was only by very close and invasive investigation methods that the real condition of the structure became apparent. It clearly highlighted the potential for issues in properties with significant chimney structures.

The discovery of the serious instability of this chimney stack despite outward appearances has since informed the Trust's approach to surveying all its historic chimney stacks.

Old Serpentine Works, Boswednack,
Cornwall

Figure 2.12 Completed works at the Old Serpentine Works. © National Trust.

The structure is a small stone building, originally a house but used as a handicraft centre for producing serpentine stone goods during the 1960s and 1970s. Prior to this it was reputedly used as a school house. Its construction date is unknown but is certainly well before 1824 when it was recorded on a land survey map of Boswednack Manor by Charles Moody.

The building is of a type rarely found beyond the parishes of Zennor, Towednack and Morvah in West Penwith. They are rectangular, single-storey granite-built houses with low-pitched slate roofs dating to the 17th or 18th centuries. Typical features include protruding kneeler stones to support the roof, a fireplace at one of the gable ends and small window openings. These buildings are some of the very few single-storey dwellings to survive in Cornwall, and possess a vernacular architecture unique to this part of West Penwith. They are similar in character to croft-type dwellings in Scotland, Ireland, Brittany and Scilly.

When this granite building was originally constructed as a farmhouse, it would have measured approximately 14.5 m externally in length by 5 m in width. Between 1880 and 1907, the southern half of the building was demolished when the road through the hamlet was widened, and the southern gable end was rebuilt. Now the building measures approximately 6.4 m by 5 m in plan. The walls are approximately 0.75 m thick. The gable-end walls stand to a height of approximately 3 m which is almost certainly the original height of the building.

In general terms, the walls of the building are constructed from granite facing blocks both externally and internally, with a granite rubble core bonded with earth mortar. There

Figure 2.13 The collapsing asbestos roof prior to commencement of work. © National Trust.

Figure 2.14 Building interior displaying staining from water ingress. © National Trust.

is evidence for repointing with lime mortar, and internal patchy remains of lime plaster and white paint. More than half of the floor area has been concreted over but a section of the earlier (probably original) flooring is visible, comprising rounded granite cobbles apparently unbonded. The concrete areas of the floor are evidence of the building's recent use as a serpentine works, as are a concrete machine base and associated iron support (possibly for a lathe) secured to the floor at the southern end of the building.

CONSERVATION PRINCIPLES APPLIED

Cornwall Archaeological Unit was commissioned by the Trust to carry out a historic building survey of the Serpentine Works. A specification for work to be carried out on the building was collated by the Trust. It set out work that was designed to renovate and return the building to its original form of construction, ensuring as much of the original fabric was maintained by completing repairs in a traditional manner. The main stonemasonry task was to make good the area of wall collapse to the front elevation and corner detailing, matching the stone with locally sourced material. The niches were found through research to be original features and it was recommended that they should not be infilled. Similar niches present on the north-east and south-east exterior elevations also remained unfilled. In addition, the internal walls were cleaned down and defective areas made good and decorated with limewash mix. A further archaeological record was made of any features that became evident after the walls had been cleaned down.

Figure 2.15 Work in progress, showing the niches to the right-hand side. © National Trust.

Figure 2.16 Side elevation. © National Trust.

Figure 2.17 End elevation showing extensive niche features. © National Trust.

Figure 2.18 Detail of head of wall and gable. © National Trust.

Figure 2.19 Completed end elevation. © National Trust.

REPAIR SPECIFICATION FOR WALLS AT OLD SERPENTINE WORKS

Note: Sections 1–2.2 omitted.

Section 2: Description of materials and workmanship
2.3 MASONRY

Materials

2.3.1 **Lime** shall be as stated in the Schedule of Works.

2.3.2 **Sand** shall be clean sharp pit sand to BS 1200, free of loam, dust, salt or organic matter, unless otherwise stated in the Schedule of Works.

2.3.3 **Water** shall be clean fresh water and free from all harmful matter.

2.3.4 **Wall ties** are to be stainless steel.

2.3.5 **Mortars** are to be as specified in the Schedule of Works.

Storage of materials

2.3.6 Lime shall be delivered in bags and stored clear of the ground in a dry weather-tight place and shall be used in order of delivery.

2.3.7 Sand shall be stored in a separate stockpile from any other materials and on a hard clean surface.

Workmanship

Generally

2.3.8 Workmanship shall generally be in accordance with CP 121 Part 1.

2.3.9 All perpends are to be kept true, square and in face work plumb with the perpends below and above, joints are to be of uniform thickness. The whole is to be properly bonded.

Protection

2.3.10 No masonry work is to be executed in frosty weather or when the temperature is 5°C or below. New masonry work is to be protected against adverse weather conditions.

Bonding new work to old

2.3.11 Where new masonry work is built up to or against existing, it shall be properly toothed and bonded.

2.3.12 Masonry work in walls or partitions abutting old work whether at right angles or in the same plane shall be bonded in alternate courses to pockets or toothings cut into the old work.

2.3.13 All new and previously laid masonry to be well wetted before use to prevent undue suction of moisture from the mortar. Ensure over-wetting does not occur.

Facing up old walls

2.3.14   Where brickwork, block work or stonework (walls, breasts, piers, etc.) has been pulled down the old work remaining shall be faced up. Work which is to be covered shall be roughly faced up, all holes or recesses being cut out square and whole bricks or half bricks being inserted so that the finished wall face is structurally sound and level enough for the covering to be applied.

2.3.15   The new work is to be tied back with headers or wall ties, as described before, according to bond.

Quoining up jambs and reveals

2.3.16   Where new openings are formed or existing openings adapted, jambs, reveals and the like are to be quoined up with new bricks, blocks or stonework as appropriate, bonded to match existing and cut toothed and bonded at the junction with old work.

Building in lintels

2.3.17   Pre-cast concrete, steel or other pre-formed lintels for openings shall have a minimum bearing on each side of the opening of 150 mm, unless otherwise stated. Any packing up and levelling under the bearing ends of lintels shall be in non-crushable materials, having regard to the weight to be supported.

2.3.18   Filling in and making good above and around lintels shall be carried out and mortar allowed to adequately set before needling or shoring is removed.

Repointing

2.3.19   Repointing old face work shall include taking out joints of all loose mortar and repointing in accordance with the Schedule of Works.

Key for plastering, etc.

2.3.20   Brick, block, stone and concrete surfaces shall be keyed for plastering or rendering as required by raking out joints as required. All dust and debris to be removed and well wetted before plastering.

2.4   AGGREGATES GENERAL

2.4.1   The coarse and fine aggregates shall comply in all respects with BS 882 and BS 1201, but the contract administrator reserves the right to reject any aggregate which, in his/her opinion, is unsuitable in any respect, notwithstanding its apparent compliance with BS 882 or BS 1201.

Fine aggregate

2.4.2   Naturally occurring sand complying with BS 882. Very coarse or very fine gap gradings are to be avoided. Sand to be free from soft materials, such as soft sandstone, limestone or coal. No crushed fines to be used.

2.4.3   Aggregates with a drying shrinkage greater than 0.06% are not to be used.

Coarse aggregates

2.4.4 Naturally occurring aggregates with gradings complying with BS 882. Soft sandstone and limestone aggregates with drying shrinkage greater than 0.06% are not to be used.

2.4.5 Flaky material will not be accepted. ALL IN or RAISED ballast shall not be used.

2.46 The maximum size of coarse aggregate to be used for the various mixes shall be 20 mm and maximum size 40 mm for mass concrete foundations.

Hardcore

2.4.6 Hardcore is to consist of hard brick, stone or ballast of approved quality, broken to pass through 75 mm ring with smaller materials to fill voids. Blinding for hardcore is to be sand and is to be laid to the specified thickness.

Reinforcement

2.4.7 Mild and high tensile steel bar reinforcement shall be to BS 3359 and mesh fabric reinforcement shall be to BS 4483. All shall be clean and free from paint, grease, millscale, loose rust or dirt and the works shall include for providing all necessary distance block, rings, ordinary spacers and No. 16 SWG tying and binding wire. The works will include for all bending, cutting and hooking, as required.

Storage of aggregate

2.4.8 Aggregate shall be stored on clean hard surfaces so as to prevent all contamination and admixture with foreign materials and to facilitate the drainage of the water. The fine aggregate shall be kept separate from the coarse aggregate.

Section 3: Schedule of detailed works

3.1 Masonry

Stonework consolidation

3.1.1 Clean up, rebuild and consolidate the collapsed area of wall to the north-west elevation to match the existing style of stonework.

3.1.2 Repairs are needed throughout the building and to replace missing stones.

Allow for localised repairs throughout the wall elevations, the floor and capping of the walls to the top of the jack rafters. Retain the niches to the elevations. Include for cleaning out and infilling the void formed by the old flue through the rear wall. Allow for the sourcing and fixing of matching stonework to fill voids in the areas where stones are missing. Bed all in lime mortar.

3.1.3 Provide and lay a lime concrete capping to the gable walls of the building to encase the purlin ends and to provide a bed for the slates. Mix to be 1:2:4 NHL 3.5 hydraulic lime:sand:aggregate.

Window W1

3.1.4 Window W1 is to be reopened to its original size, approximately 400 mm high × 550 mm wide as indicated internally. Utilise the removed stonework in the consolidation works.

3.1.5 Leave the window opening ready for the fixing of the new window.

3.2 Repointing

Test panels

3.1.6 Provide at least a 1m² test panel of pointing to the rear elevation of the building for approval by the client before the remaining pointing proceeds.

3.1.7 This must be prepared and left to go off for at least one week before the inspection is carried out and must be protected from drying out too quickly by dampening and covering with damp hessian sacking for at least 90 hours.

3.1.8 The panel to include the following mixes and finishes.
- 2½ parts NHL 3.5 to 3 parts Doble yellow sand to 3 parts sieved Geevor grit.
- Stipple finish with a churn brush to draw the aggregate to the surface and slightly recessed from the arrises.
- Hessian sack rubbed finish to be flush or slightly recessed from the arrises.

3.1.9 Allow for the retention of the approved test panel until near completion of the repointing works, at which time it should be hacked out carefully and repointed in the approved mix and style.

Preparation

3.1.10 Ensure the backs of the cleaned out joints are square and clear of any loose debris.

Damping

3.1.11 The joints must be thoroughly dampened with enough time left for the stone faces to dry in order to prevent smearing of the mortar onto the stone face. The mortar should be as dry as is practicable to point with.

Pointing

3.1.12 Commence pointing at the top of the wall and work down. Use a suitably sized pointing key or metal spatula to force the mortar fully into the joints from a hawk. Joints over 25 mm deep to be dubbed out with mortar at least one week prior to the final pointing.

Finishing

3.1.13 Allow for brush tamping/sack rubbing of the joints when the mortar is green hard.

Protection

3.1.14 Maintain full protection from drying winds, direct sunshine, rain and frost with hessian sacking or other approved coverings. Include for mist spraying of the work to prevent rapid drying out of the mortar and ensure it is not over-wetted.

Mortar mix

3.1.15   For tendering purposes assume the following mix: 2½ parts NHL 3.5 to 3 parts
well-washed Doble sand to 3 parts sieved Geevor grit.

Areas to be repointed

3.1.16   Repoint all areas previously hacked out to the walls and floor.

Specification author: National Trust

COMMENTS ON RESULTS

The conservation work was specified and procured by a Trust building surveyor. The contract
was carried out by a local building contractor based in St Just.

Damaging ivy and other vegetation have been removed from the building. Areas of
collapsing stonework have been carefully rebuilt and the whole building has been repointed
inside and out with the inside having a limewash finish. The external joinery has been
replaced in the local style.

The asbestos roofing was removed. The aim of the project was to return the building to
its original form of construction and to a large extent this was done. However, due to financial restrictions, the roof was not recovered with scantle slate but instead with corrugated
iron. This makes the building useable and it will be easy to remove and replace with scantle
slate when funds permit.

The works have turned a ruinous building that was in a state of collapse back into a
useable weather-tight condition. It has been let to a local person who uses it as a store.

Chapter 3

Timber repairs

There is a long tradition of building with timber in Britain, with many methods and tools perfected millennia ago. The process of specifying timber work is more modern in comparison but has changed little from 1898 when the first trade-based clauses were published in F.W. Macy's *Specification in Detail*.

Two small and quite unusual structures are the subject of the case studies in this chapter: the gatehouse at Brockhampton and the Saddlescombe donkey wheel. Despite their size and uniqueness, their specifications exhibit many common aspects of the repair of historic timber structures. Traditional methods were chosen for these repairs and not only for conservation reasons; regardless of principles, there were the best and, in many cases, still the only methods that work.

Conservation specifications will typically need to address both hardwood framing methods and the various internal softwood uses. Skilled carpenters and joiners may fairly readily adapt to conservation work as the methods and construction will be familiar to them, even if they do not use them often on modern buildings. Not all skilled craftsman will have developed their experience and understanding, however, and the specification should anticipate this.

The Saddlescombe specification is clear about the traditional jointing methods required (in this case, peg fixing) and reminds the craftsmen that the moisture content of timber used for repair must match that of the original wood. Equally, that the conversion of replacement timber should 'follow the conversion from the log dimensions of the member to be replaced' so that moisture movements match as much as possible. These points would be obvious to some contractors but not to all.

The Brockhampton specification takes this one step further by indicating the wider implications of this technique and pointing out that draw boring 'requires the erection and dismantling of frame members twice before final pegging takes place'. This is standard best practice for this type of construction.

A typical approach in specifications is to keep the Schedule of Works simple by referring to detail in drawings and to sections on materials and workmanship. Following this approach in conservation specifications has the added benefit of letting the Schedule of Works focus on these prescriptive instructions required for traditional methods.

A conservation specification is predominantly prescriptive but it will also mix in performance and benchmarking. The Saddlescombe specification names the manufacturer of the nails to be used, and the Brockhampton specification often makes reference to British Standards. In some cases, the British Standards are used as a baseline from which to build up a prescriptive instruction.

These specifications make good use of drawings; typically, drawings better indicate assembly procedure and specifications better describe quality but there is often overlap. Some practitioners will see drawings as the primary source of construction information. Notes on the drawings will then be important but there will always be limited space. They can be annotated with co-ordinated project information references from the specification, of course.

The Gatehouse, Lower Brockhampton
House, near Bromyard, Herefordshire

Figure 3.1 Brockhampton
Gatehouse. © National Trust Images/
Robert Morris.

CONTEXT

Lower Brockhampton House is a base-cruck hall with spere-truss built within a wet moat
during the early years of the 15th century. To the south of the house, and erected circa
1600, stands a small gatehouse. It originally straddled the moat but its western wall now
stands on the 19th-century dam which closed the moat.

The gatehouse is a two-bay, close-studded symmetrical timber frame at first floor, jettied
on all four sides over a two-bay asymmetrical ground floor. Its heavy oak door is carried
from the central frame. The structure is top-heavy, out of line and dependent upon four
bressumers (long supporting beams) and a strong, rigid frame.

Both the house and the gatehouse underwent change and deterioration until they were
restored in the mid-19th century. This repair of the gatehouse, although partly driven by a
fashion for the 'picturesque', did attempt to address a fundamental structural inadequacy

that was causing a gradual descent to the east. Corbel brackets were nailed beneath the jettied eastern floor joists and a pair of braces installed into the moat walls to support the eastern wall frame. It was during these works that all infill panels and some timber-frame members were replaced with lath and plaster.

The Brockhampton Estate, including the house and its gatehouse, was given to the Trust in 1946 by which time both buildings were in poor condition. A major repair scheme was undertaken during the 1950s to arrest the main problems. This included reconstruction of the entire upper south wall frame of the gatehouse on the existing bressumer, with cover patches over the mortises and a great deal of patch repair that helped slow further decay.

In 1991 the Trust commissioned inspections of both buildings and an adjacent ruined chapel. Following advice upon their structural condition, the Trust undertook a major tim-ber-frame repair to the house during 1995–6, with the assistance of Historic England. It was at this time that a conservation and repair philosophy was developed for both the house and future gatehouse repairs. This was heavily influenced by the importance of the mid-19th century 'picturesque' restoration, which had become part of the buildings' style.

Because of its configuration and construction, the gatehouse has always been a fragile structure. The steep staircase, which gives access to the first floor, was cut through the first-floor joists and spine beam, compromising the structural integrity of the east side of the floor frame. As a consequence, the load from the eastern wall frame, especially the centre-truss storey post, caused the jettied joists to 'droop' and the east bressumer to sag at its centre. Compounding this distortion, the cill beam of the eastern wall frame spanning the moat also deflected so that the entire structure began, probably quite soon after com-pletion, to capsize to the east. The deformation was progressive, causing the floor joists and beams to crush and break at their bearings on the frames beneath so that the upper storey moved further out of alignment, eventually far enough to allow rainwater from the unguttered west roof slope to be shed directly onto the top of the west bressumer with its close-stud frame mortises uppermost. Inevitably, the east bressumer broke and the west bressumer became seriously decayed, as did the southern bressumer, the most exposed to the prevailing wind direction.

By 1998 the gatehouse's upper floor had to be closed off when its stability was further compromised by the splitting of the post head jowls of the central truss caused by the distortion of the eastern side. Repairs were set in hand for 2000.

CONSERVATION PRINCIPLES APPLIED

Although it was clear that the gatehouse required major repair to ensure its survival, its asymmetry and appearance could not be overly 'corrected' as these were deemed to be part of the building's significance. After analysis of the structural behaviour of the frame and its defects, it was initially proposed to replace three of the bressumers to ensure the long-term stability of the gatehouse. Because of their size, over 3 m in length and 450 mm square, new timber would be costly and difficult to source. A compromise was reached, whereby only the east and south beams would be replaced because of their extreme states of decay and previous alteration; the western beam would be repaired.

Figure 3.2 Superstructure lifted revealing condition of original tenons. Note bracing used to preserve alignment during lifting. © Caroe and Partners, Architects.

Figure 3.3 Internal jacking apparatus for lifting superstructure. © Caroe and Partners, Architects.

Figure 3.4 External jacking apparatus for lifting superstructure. © Caroe and Partners, Architects.

In conjunction, the floor joists and beams would be retained *in situ* and strengthened individually with steel flitches. The tendency for the floor frame to lift at its centre was overcome by introduction of a tie rod within the thickness of the ground-floor central truss between cill and spine beams. To achieve this, the entire upper frame, without its stone-tile roof covering, would have to be lifted whilst the floor frame was repaired and then lowered back into place on to the repaired timbers. This kind of procedure is not common as it is fraught with difficulty and attendant risk of structural damage to the retained and lifted elements.

Structural engineers designed the lifting mechanism and scaffolding framework to support it. Twenty car jacks were used as supports beneath the wall plates and tie-beams, and were sufficient to raise the superstructure 300 mm from the bressumers after all of the pegs had been removed and the joints eased. In addition, a coffer dam had to be set into the moat to allow the section beneath the gatehouse to be drained but kept wet to avoid shrinkage of the puddle clay. A temporary footbridge was placed over the moat to provide access to the house for visitors and to provide a grandstand for inspection of the works in progress.

There was concern that the repairs would necessitate a considerable amount of realignment to ensure stability of the timber frame and to avoid unacceptably high asymmetrical loads on its joints and centre truss posts. It was agreed, however, to limit correction of the

Figure 3.5 Superstructure lowered back on to new bressumers. Note the replication of the downward curvature of the east bressumer. © Caroe and Partners, Architects.

tilt by forming the east bressumer with a downward curve throughout its length and by the use of slip tenons to joint the already slightly shortened wall studs to it. Although the tilt has been redeemed by 100 mm, the alteration has been disguised by this method. To protect the west bressumer from roof water, a discreet gutter has been incorporated on the west roof slope.

REPAIR SPECIFICATION FOR TIMBER FRAME OF THE GATEHOUSE, BROCKHAMPTON

Notes

1. Numbering within this specification relates to the National Building Specification.

2. Section 1 omitted.

Section 2: Materials and workmanship

C10   Dismantling structures

165   TIMBER FRAME DISMANTLING

The contractor should include in his rates for taking down or dismantling of timber framing for the taking out and discarding of unwanted timbers; taking out straps, bolts and the like (excluding those in existing repairs which are not to be disturbed); taking out all timbers necessary to enable repairs to be carried out for new work or repaired members to be inserted and for making good all work disturbed.

The contractors must also allow for the fact that timber frames need to be dismantled and repaired in stages to ensure that as little as practicable of the framing is dismantled at any one time.

Prior to dismantling, existing joints are to have their pegs removed by hammering out from the lower face; drilling out will not be permitted.

G20 Carpentry/timber framing/first fixing

To be read with Preliminaries/General conditions.

Type(s) of timber

320 HARDWOOD FOR JOINERY ITEMS

Quality of timber: BS1186: Parts 1 and 3.
Species: European Oak.
Class: CSH.
Moisture content at time of fixing: 12% to 14%.

325 OAK FOR TIMBER FRAMING

Type: Unseasoned native oak from selected timbers or standing trees.
Supplier: To contractor's choice subject to architect's agreement.
Finish: Sawn.
Conversion: All required members to be obtained from logs of the least diameter yielding correct finished cross-sectional size. Small areas of sapwood permitted but no sapwood to be exposed externally or covered internally.
Members greater than 200 mm × 200 mm to be boxed heart.
Members less than 150 mm in their lesser cross-sectional dimension to be heart sawn.

Workmanship generally

400 BASIC WORKMANSHIP

Comply with the clauses of BS 8000: Part 5 which are relevant to this section.

405 ACCURACY

Notwithstanding BS8000: Part 5, clause 3.2.1, comply with Preliminaries clause A33/340 and any required critical dimensions given in the specification or on the drawings.

410 CROSS-SECTION DIMENSIONS OF TIMBER

Shown on drawings are basic sizes unless stated otherwise. Maximum permitted deviations from basic sizes to be as stated in BS5450 for hardwoods.

415 CROSS-SECTION CHARACTERISTICS OF OAK FOR TIMBER FRAMING

Normal method of conversion to be as summary below except where varied by Repair Schedule:

Beams (floor-, sill-, cross-)	Boxed heart
Purlins	Boxed heart

Plates (wall-, arcade-collar-, sole-,)	Boxed heart or quarter-sawn
Ridge	Boxed heart or quarter-sawn
Posts	Boxed heart or halved
Tie-beams	Halved
Girdings	Halved
Bressumers	Halved or quarter-sawn
Studs	Halved
Braces	Halved
Brackets	Halved
Principals	Halved
Collars	Halved
Struts	Halved
Joists	Halved or quarter-sawn
Rafters	Boxed heart poles squared on three sides
Barge boards	Quarter-sawn

420 SHAPING OF OAK FOR TIMBER FRAMING

Straight members: Sawn or axed on sides other than any exposed heart face.
Curved members: Sawn on faces not curved but axed or adzed on curved faces so that final shape conforms to natural grain.

Shaped members: Members such as jowled heads of posts, tie-beams, etc., to be axed or adzed to shape in accordance with natural grain. Non-shaped faces to be sawn.

430 SELECTION AND USE OF TIMBER

Do not use timber members which are damaged, crushed or split beyond the limits permitted by their grading.

Ensure that notches and holes are not so positioned in relation to knots or other defects that the strength of members will be reduced.

Do not use scarf joints, finger joints or splice plates other than as scheduled to timber-frame repairs.

440 PROCESSING TREATED TIMBER

Carry out as much cutting and machining as possible before treatment.

Retreat all treated timber which is sawn along the length, ploughed, thicknessed, planed or otherwise extensively processed.

Treat timber surfaces exposed by minor cutting and drilling with two flood coats of a solution recommended for the purpose by main treatment solution manufacturer.

510 PROTECTION

Keep timber dry and do not overstress, distort or disfigure sections or components during transit, storage, lifting, erection or fixing.

Store timber and components under cover, clear of the ground and with good ventilation. Support on regularly spaced, level bearers on a dry, firm base. Open pile to ensure free movement of air through the stack.

Arrange sequence of construction and cover timber as necessary during and after erection to ensure that specified moisture content is not exceeded.

521 TREAT

Treat exposed end grain of all preservative-treated timbers with same preservative brush applied on site.

540 CLEAR FINISHES

Structural timber which is to be clear finished to be kept clean and first coat of specified finish applied before delivery to site.

550 EXPOSED TIMBER

Prevent damage to and marking of surfaces and arrises of planed structural timber which will be exposed to view in completed work.

Repairing timber framing

610 TIMBER REPAIR JOINTS

Decayed existing framing members to have new oak sections connected by joints designed for complete rigidity so that when finished the member again acts as a single component.

Joints to be formed as scheduled for timber-frame repairs.

Joints to be both pegged and glued.

Pegs: Formed as clause 670; inserted in skewed alignment, approx. 150 mm CCS, avoiding continuous grain. Pegs to be cut off flush at both ends after shrinkage has taken place.

Adhesive: Synthetic resin to BS 1204: type WBP and to have gap-filling character-istics as 'Cascamite', Humbrol Ltd., Marfleet, Hull, or other similar approved.

620 METAL REPAIR JOINTS

Fractured existing framing members, where to be retained, to be strapped on both faces and bolted through sufficient to prevent further distortion of the damaged member rather than forcing it back to its original form or position.

Plates: Non-ferrous or austenitic stainless steel of size and thickness as scheduled for timber-frame repairs.

Bolts: Non-ferrous or austenitic stainless steel of diameter and number as scheduled for timber-frame repairs.

630 FILLED TIMBERS

Partially decayed existing framing members where to be retained to have shakes, knots, etc. filled with a thixotropic mixture of adhesive, as clause 610, and sawdust.

Cement mixtures are not to be used.

Filled areas to be cleaned of dirt, loose material, etc., prior to carrying out filling operations.

Where filling is to timbers adjoining construction joints, filler is to be separated from non-filled member by aluminium foil which is to be removed when filler has set.

640 BLOCKED TIMBERS

Concealed mortise pockets, redundant peg holes, etc. are to be blocked with oak sections glued, but not pegged, into position as clause 610.

Mortise pockets are to be cleaned out preparatory to blocking and peg holes are to be drilled to approx. 35 mm diameter.

Blockings are to be cut to be a tight fit within the openings with the grain of the blockings following that of the timber member.

Jointing timber

650 JOINTING/FIXING GENERALLY

Where not specified otherwise, select fixing and jointing methods and types, sizes and spacings of fastenings in compliance with section 220. Fastenings to comply with relevant British Standards.

670 TIMBER FRAMING CONSTRUCTION JOINTS

New timber member to timber member joints or reassembly of existing similar joints to be carried out as a frame lies horizontally before rearing to the vertical or fitted in sequence as the frame is vertically assembled.

New tenons to construction joints to allow slight play but the shoulders of tenoned timbers must fit tightly to the mortised member.

All construction joints to be pegged with only internal projecting ends cut to approx. 35 mm clear of the frame face. External projecting ends not to be cut unless instructed by the architect at the end of the defects liability period.

All new construction joints to be draw-bored so that holes are staggered to give tightening of the joint as the peg is driven in. Tenoned members must be drilled separately from mortised members.

Pegs: Oak, hexagonal or roughly squared, tapered and formed by cleaving and not sawing. Length to be thickness of member to be penetrated plus 75 mm. Pegs to be hammered in from the 'upper face' of the frame.

675 REMOVING EXISTING PEGS

Existing pegs to be removed by hammering out.

No drilling will be permitted.

680 BOLTED JOINTS

Locate holes accurately and drill to diameters as close as practical to the nominal bolt diameter and not more than 2 mm larger.

Place washers under all bolt heads and nuts which bear directly on timber. Use spring washers in locations which will be hidden or inaccessible in the completed building.

Tighten bolts so that washers just bite the surface of the timber and at least one complete thread protrudes from the nut.

Check at agreed regular intervals up to Practical Completion and tighten as necessary to prevent slackening of joints.

690 ROOFING MEMBRANE

Manufacture and type: Roofing membrane, 'Type 234' manufactured by British Sisalkraft Ltd, Strood, Kent, or other equal approved by the contract administrator. Fixing: Stainless steel staples at 100 mm centres.

Erection and installation

770 ADDITIONAL SUPPORTS

Where not shown on drawings, position and fix additional studs, noggings or battens for appliances, fixtures, edges of sheets, etc., in accordance with manufacturers' recommendations.

All additional studs, noggings or battens to be of adequate size and have the same treatment, if any, as adjacent timber supports.

800 ASSEMBLING TIMBER FRAMING

Re-erecting or reassembly of timber framing is to allow for incorporating existing members, repaired members and new members and for pegging construction joints, for all hoisting. jacking, supporting, etc., to provide repaired frames as scheduled.

The contractor is to note that the procedure of marking and boring necessitated by draw-boring requires the erection and dismantling of frame members twice before final pegging of joints takes place.

The contractor should also note that difficulties will be encountered in working with frames that will generally not be square, plumb and true.

Section 3: Schedule of Works

44 Dismantling

441 GROUND FLOOR BOARDING

Identify each board to be lifted with aluminium tags stamped with identity numbers; nail tags to board soffits.

Carefully punch out all existing nail fixings from floor boards Nos. 1, 2, 3, 11, 12, 13, 14, 15, 16, 17, 18 and 19.

Lift boards with softwood wedges and levers to avoid damage to 6 No. existing boards to be reused; set aside under dry ventilated cover. Discard remainder of removed boards and cart from site.

Allow for removal of inserted brick and stone joist end supports as work proceeds; cart removed material from site.

442 FIRST-FLOOR BOARDING

Identify all floor boards with aluminium tags as described in clause 441.

Carefully punch out existing nail fixings and lift boards with softwood wedges and levers and set aside under dry ventilated cover for reuse.

Remove all existing nail fixings from joists over all floor frame.

443 FIRST-FLOOR INFILL PANELS

Carefully take out lath-and-plaster infill panels to each elevation at first-floor level; discard removed materials and cart from site.

Clean down reveals to panel-surround framing including removal of any remaining nail fixings.

444 FRAME LIFTING

Install scaffolding/lifting apparatus beneath tie-beams and wallplates as agreed with structural engineer under clause 421.

Carefully remove all existing pegs by driving out of wall frame, studs, corner posts and intermediate truss posts and bressumers.

Notify architect and structural engineer of detailed daily programme for lifting operation and arrange for them to be in attendance during lifting works. **DO NOT** carry out the lifting works without the attendance of the architect and structural engineer on site (C10/610).

Lift upper storey framework sufficient to allow for removal of bressumers.

NOTE: The north bressumer is to be retained in situ but the south, east and west bressumers are to be removed and either replaced or repaired. It is essential therefore that the mortise and tenon joints of the north bressumer/wall frame are undamaged but that the other three may be adjusted or severed in order to facilitate lifting.

When first-floor frame has been lifted clear of the substructure, carefully remove existing pegs by driving out from between floor joists and bressumers of west, south and east sides only.

Mark bressumers with numbered aluminium tags as described in clause 441; remove from floor frame and set aside for examination by architect and structural engineer.

45 Frame repairs – Phase one

To be undertaken whilst first-floor frame is in its dismantled and lifted state.

451 GROUND FLOOR STRUCTURE

Carefully take out lath-and-plaster infill panels, 3 No., adjacent to north-west elevation doorway; discard removed materials and cart from site.

Take out inserted timber wedges from head of frame, doorpost, doorhead and main first-floor beam.

Drill through beam 'X' and floor beam 'W', intermediate rail and drill within the thickness of the infill panels in the position shown on drawing 591/90/11. Set stainless steel threaded tie bar through holes, bar 20 mm dia. (G12/11 0).

Secure bar to the first-floor frame with a 75 mm square × 5 mm thick stainless steel plate, spring washer and nut all housed into the top of beam 'X' and similar plate and nut fixing on the soffit of beam 'W' not housed in. Tighten tie bar until floor frame is reset into correct relationship to post head and repeg into existing holes (G20/61 0).

452 FIRST-FLOOR STRUCTURE

Carefully cut out vertical slots to each end of both main floor beams in positions shown on drawing 591/90/06, slots sized 100 mm deep × 6 mm wide.

Insert new stainless steel flitch plates into slots, plates 5 mm × 75 mm × 1000 mm long (G12/11 0).

Secure flitch plates with 12 mm diameter stainless steel threaded rods bolted through beams and flitch plates, 13 No. bolts per plate.

Set bolt ends and nuts below surfaces of beams and fit oak cover pellets; saw pellets flush to beam faces.

453 FIRST-FLOOR JOISTS

Insert new sawn oak joists over existing corbel brackets to complete length of east elevation, joists 175 mm deep × 125 mm wide × approx. 550 mm long (G12/110); profile joist depths to varying wedge shapes to give level upper surface as shown on drawing 591/90/06 at details 'B' and 'C'.

Peg new joist sections to existing joists following refitting of bressumer as later described (G20/670).

Provisional sum included elsewhere for renewal of joist wall/frame pegs and replacement of defective timbers.

454 FIRST-FLOOR JOIST STRAPS

Fix stainless steel straps across broken corbelled joists to east elevation, straps 25 × 5 mm × 600 mm long (G12/110); house straps into upper surfaces of joists; fix with stainless steel screws.

Allow for 6 No. joists to be strapped.

455 EXISTING STAIRCASE STRAP

Clean off rust deposits from existing steel strap within first floor at head of staircase; prepare for decoration.

Apply 2 No. coats preservative metal paint to strap (M60/120).

456 FIRST-FLOOR WALL FRAME BRESSUMERS

On completion of realignment of first-floor frame to continuous levels at perimeter, carry out work in bressumer replacement. Replacement to be either:

i. Provide and fix 3 No. new oak bressumers to north-west, south-west and south-east wall frames, bressumers from boxed heart sawn timber to finished size 230 mm × 175 mm complete with mouldings to match the original and mortises to suit studs of wall frames (G20/325).

or

ii. Repair existing 3 No. oak bressumers by cutting away all decayed timber to leave sound outer moulded face; laminate moulded faces to new 100 mm × 80 mm boxed heart sawn oak beam (G20/325); lamination to be with resin glue and stainless steel countersunk and pelleted screws.

Include mortises as before to suit existing wall frames.

In both cases, include for cutting chamfer to top external edge to each bressumer to shed water from joint.

Repair retained bressumer to north-east elevation by careful removal of existing face repair timber; cleaning out and checking mortise pockets behind removed timber and refixing of timber back into position.

457 FIRST-FLOOR WALL FRAME TIMBERS

Carefully take out decayed and damaged studs and rails within wall frame as follows:

4 No. studs to south-west elevation

1 No. rail to south-west elevation

1 No. stud to south-east elevation

Cut new sawn oak studs and rails, sizes as scheduled on drawings 591/90/09 and 10 (G20/325); form tenons to ends of timbers and mortise studs for rail.

Peg new timbers into position.

458 GROUND FLOOR WALLPLATE

Apply preservative treatment to existing cill beams and floor joist ends exposed by lifting of floor boards (G20/440); apply preservative by brush in 2 No. coats.

Fix new sawn oak wallplates to inner faces of existing soleplates to south-east and north-west elevations, size 100 × 50 mm (G20/325); fix with stainless steel coach screws; include for cutting housings to plate at joist intersections.

Set Code 5 lead DPC tray over stone plinth base and dress below and behind new wallplates as shown on drawing 591/90/06 (H71/320).

Fix stainless steel joist hangers to wallplate and nail to plate and to joists with stainless steel nails.

459 GROUND FLOOR BOARDS

Take existing retained boards from store and refix into numbered positions; fix boards with cut nails (K20/370).

Allow for planing edges to refixed boards to give tight, cramped fit to new replacement boards.

Fix new sawn oak boards to positions of discarded boards, size 30 mm thick × width as existing (K20/120); fix boards with cut nails; plane edges to allow for tight, cramped fixing.

Include for 6 No. new boards.

4510 FIRST-FLOOR BOARDS

Take existing removed boards from store and refix into numbered positions; fix boards with cut nails (K20/370).

Provisional sum included elsewhere for planing of board edges found to be necessary to achieve tight, cramped fit between boards.

4511 FRAME LOWERING

Notify architect and structural engineer and arrange for attendance as clause 444. DO NOT carry out the lowering works without the attendance of the architect and structural engineer on site (C10/610).

Carefully lower upper framework onto new or repaired bressumers, locating each stud, corner post and intermediate post into prepared mortises as detailed and repeg with new riven oak pegs in original and new peg holes.

NOTE: *To east elevation in particular because of the deformation of the wall frame, wallplates and original bressumers, adjustment of the height of the first-floor studs by removal of decayed timber tenons and reformation of new tenons will be necessary.*

It is intended that no deformation greater than 75 mm will be perpetuated on the south-east wall frame.

Include for cutting existing timbers to form new tenons to achieve required alignment (G20/670).

Allow for 15 No. new tenons.

46 Frame repairs – Phase two

461 NORTH-EAST TRUSS

Carefully cut back faces of frame members to allow for face patch repairs as follows:
  1 No. end to truss tie-beam on inner east face
  1 No. top to frame post on inner east face.

Cut slot to truss principal at east end, slot 6 mm wide.

Carefully cut and take out sections of frame members as follows:

1 No. wallplate to south-east wall frame at truss junction.

Fabricate new stainless steel T-shaped flitch plate to profile shown on drawing 591/90/08, plate 5 mm thick (G12110).

Cut new sawn oak wallplate, size as scheduled on drawing 591/90/08 (G20/325); form tenoned bridle to end of new wallplate, dovetail housing for tie-beam and mortise for post.

Peg wallplate into position.

Cut new sawn oak face patches for tie-beam and frame post, sizes as scheduled on drawing 591/90/08 (G20/325).

Fix flitch plates to truss tie-beam, truss principal and post head and sandwich with new oak face patches; bolt through timbers and plate with 12 mm dia. stainless steel threaded rods, 19 No.

Set bolt ends and nuts below surfaces of timbers and fit oak cover pellets; saw pellets flush with timber faces.

Replace existing pegs within adjoining timber frame members where removed to carry out works; new pegs to be riven oak (G20/670).

462   SOUTH-EAST WALL FRAME

Carefully cut back faces of frame members to allow for face patch repairs as follows:

1 No. angle brace end on inner face

2 No. angle brace ends on outer face

4 No. stud ends on outer face.

Carefully remove previous face board repairs and trim back timber to provide sound surface for face patch repairs as follows:

1 No. corner post at base to both outer faces.

Cut new sawn oak patches, min. 50 mm thick ×face sizes as scheduled on drawings 591/90/09 and 10 (G20/325).

Peg and glue new timbers onto cut back sections; ensure that new faces to base of corner post restore original moulded and water shedding profile (G20/610).

463   NORTH-EAST TRUSS

Carefully remove previous face board repair and clean down board and backing timber for inspection as follows:

1 No. tie-beam at centre of truss.

Allow for inspection by architect and structural engineer.

Provisional sum included elsewhere for repairs to truss deemed necessary following uncovering.

Refix existing face board; peg board into position.

Take out all existing pegs to joints between tie-beam and tenoned studs and braces; repeg in existing holes with new riven oak pegs.

464 INTERMEDIATE WALL FRAME POSTS

Prepare heads of intermediate truss posts for resin repair by withdrawing pegs to corbel braces, taking out braces, tag marking and setting aside for reuse.

Wrap post heads with non-slip membranes secured by tape and infill mortises and peg holes with polystyrene.

Cross-drill post heads and insert 6 No. 12 mm dia. stainless steel threaded dowels, infill cavity in each post head with resin (G201710); extent of filling and location of dowels to be as scheduled on drawing 591/90/11.

On completion of void fill, remove wrapping and polystyrene from mortises. Insert arched braces and repeg.

465 WALL FRAME WALLPLATES

Fabricate 2 No. new mild steel angle reinforcement plates, size 100 × 75 × 5 mm × 1300 mm long each (G12111 0); drill each plate for 7 No. vertical and 7 No. horizontal bolted fixings.

Fix plates over wallplates to south-east and north-west wall frames, centre plates on intermediate trusses; bolt plates through wallplates with 10 mm dia. stainless steel threaded rods, washers and nuts.

Set bolt ends and nuts below inner faces of wall plates and fit oak cover pellets; saw pellets flush to wallplate faces.

Paint plates prior to fixing in position with 2 No. coats of metal paint (M60/110).

49 General frame repairs

491 INFILL PANELS

To all removed or damaged infill panels as marked on drawings form new 50 × 25 mm oak subframes fixed to inner faces of all frame studs (G20/325); fix subframes with stainless steel screws.

Fix new riven oak plaster laths to subframes (M20/110).

Apply 3 No. coats hair-reinforced render to new infill panels and to damaged infill panels remove loose surface material and re-render with hair-reinforced plaster to match the new panels (M20/110).

492 TIMBER REPAIRS

Take all existing Cascamite and sawdust repairs from external elevations to timber frame and include for infilling minor shakes, cracks and cavities with hair-reinforced lime: sand plaster (M20/110).

Provisional sum included elsewhere for piecing in oak slivers to larger shakes and cavities.

Include for inspecting frames with architect following removal of existing fillings and agreeing extent of openings to be infilled with oak sections.

493 DECORATIONS

On completion of repairs to timber framework and infill panels, apply 5 No. applications of lime wash over all the external timber and plaster surfaces and to all internal plaster surfaces only (M60/140).

NOTE: All internal timber framing is to be left _undecorated_.

When lime wash has been completed, allow 7 day period and brush lime wash from timber-frame members only, using wire brushes.

Include for work at end of 12 months defects liability period in applying 3 No. further coats of lime wash and repeat the external brushing (M60/140).

410 Joinery repairs

4101 WINDOW FRAMES

Construct new wrot oak frames to original pattern but with cills projecting minimum 38 mm from the elevation (L20/11 0).

Refix frames into existing openings setting cills on new code 4 lead sub cills dressed 25 mm over face of rails (H71/210).

Seal around perimeter edges of window frames with bituminised expanding foam filler strip (L20/81 0); set filler to finish 10 mm behind faces of frames.

Treat window frames prior to fixing with 3 No. coats of raw linseed oil and decorate exposed faces following fixing with further 1 No. coat of oil (M60/150).

4102 GLAZING

Remove salvaged glazing panels from store and reglaze into new oak frames using new lead cams to suit diamond-shaped quarries (L201770).

4103 DOORS

Take existing doors from store and rehang on existing pintles. Include for formation of new phosphor bronze shims to pintle hinge or main entrance door and lubricate with copper lubricant.

To both doors treat all metal work with clear micacious oxide (M60/130).

No other decoration works to timber to be carried out.

Specification author: Ian Stainburn, Caroe and Partners, Architects

COMMENTS ON THE RESULT

Causes of the structural failure of the gatehouse were identified and understood by thorough survey and analysis. These were rectified successfully using a mixture of traditional carpentry techniques, with modern intervention only where necessary.

The historic significance of the structure was recognised, as was the requirement to maintain its popular picturesque deformation. There are still reservations about this because no designer or carpenter would replace a damaged bressumer with a curved one, nor would one normally be required for structural reasons. Maintaining the asymmetry and tilt created other difficulties; for example, the need for a gutter to be added on the west side to prevent water coursing off the roof into the mortises of the west bressumer.

Saddlescombe Donkey Wheel, Newtimber, West Sussex

CONTEXT

Saddlescombe is a downland estate with a largely intact working model farm. See page 23 for a full description of the farm.

Part of the farm is the Grade II-listed donkey wheel, thought to date from 1855 but probably with 17th-century origins, which is believed to be one of only four remaining in south-east England. The vertical wheel is some 3.6 m in diameter and is housed in a one-bay timber structure just large enough to accommodate it, built on a low flint-and-brick wall. The wheel has timber treads and a donkey, or perhaps a pony, provided the motive power by treading the boards. A chain runs around the axle and circumference of the wheel, which was used to draw water from a 1.8 m diameter well dug at least 45 m into the chalk downs, using two buckets attached to the ends of the chains (as one was raised, the other would lower). Above the well cover is a lead-lined gulley about 2.4 m long into which the water was poured before being distributed along lead pipes. Much of the building, the wheel, cover and gulley were built in oak originally, although some elm was used for the treads and

Figure 3.6 Early photograph of the donkey wheel.© National Trust.

Figure 3.7 Early photograph of the donkey wheel. © National Trust.

tank boards. Three sides of the building are clad and the fourth side is open for access. The building and wheel have received varying degrees of repair to the structural timbers and weatherboarding over the years. The treadwheel itself did not require any major work, but the axle support and the bracing across the east elevation had failed, causing the building to list and the wheel to foul the sides. Much of the previous structural repair work had been executed with softwood which was decaying prematurely. Large areas of weatherboarding had also been replaced with softwood boards which had decayed rapidly, allowing water to penetrate and further affect the structural timbers behind.

A condition survey was undertaken in 2000 by specialists in the repair and conservation of historic timber-framed buildings. The work was eventually undertaken during 2004 by local timber-frame specialists, who found that the well cover and gulley had deteriorated to the extent that they were in danger of collapse and also needed major repairs. Funding was provided by the Trust and the Department for Environment, Food and Rural Affairs. The repairs were undertaken in two stages lasting a total of four weeks.

One of the first tasks was to source local oak for the repairs and allow it to dry in the microclimate of Saddlescombe Farm. As much original timber as possible was reused and only essential repairs were undertaken. An example is the repair to the axle support which was achieved simply by inserting wedges to raise the axle sufficiently.

Repairs were undertaken with honesty and not disguised, although stainless steel bolt heads were toned down with matt black paint. Resins and metal strapping were largely

Figure 3.8 East elevation. © National Trust.

Figure 3.9 South and west elevation. © National Trust.

avoided. The opportunity to make improvements was taken where this benefited ongoing conservation; for example, the ground level at the open (east) side was altered to ensure that rainwater was directed away from the building, ventilation was incorporated into the well cover and cement mortar was replaced with lime mortar.

109

Figure 3.10 Minimal replacement of boards to south end of gulley casing. © National Trust.

Figure 3.11 Wedging and pegging of axle support member and east tie beam. © National Trust.

CONSERVATION PRINCIPLES APPLIED

The repair strategy followed the principles of the Society for the Protection of Ancient Buildings. The intention was, as far as possible, to conserve the original fabric and to undertake reversible repairs.

Figure 3.12 East cill beam, scarf repairs to right post. © National Trust.

SPECIFICATION FOR TIMBER REPAIR TO DONKEY WHEEL AT SADDLESCOMBE FARM

Notes

1. Numbering within this specification relates to the National Building Specification.

2. The original specification was transcribed and put into National Building Specification format.

Informative

- **South elevation:** A long section of oak cill has been replaced with a softwood member. This member has deteriorated and requires replacement. The studwork and brace, which connects into this member, has cut off stub tenons. The plinth wall has been rebuilt with a cement-based mortar.
- **West elevation**: This elevation has received minimal repair work in the past.
- **North elevation:** This elevation has similarities with the south elevation. The oak cill has been repaired with softwood timber that is now decaying. A short section of plinth wall has been rebuilt using fletton bricks in order to realign this wall in a perpendicular relationship with the wallplate. The bracing in the east elevation has failed to restrain the building in the original alignment.
- **East elevation:** This open elevation relies on the braced east donkey wheel axle support structure, which is positioned inside the line of the east posts to

restrain any lateral movement to this elevation. This internal support frame-work relies on an oak cill plate, which has become very decayed. This plate is presently embedded in the ground and located in between the bases of the north and south plinth wall. The failure of this cill plate has caused the central axle frame to drop and braces to become ineffective, allowing the building to lean northwards.

C01 Schedule of Work

SOUTH ELEVATION

100 REPLACEMENT OF SOFTWOOD CILL BEAM: As section G20

101 Prop elevation with braced scaffolding and needles

102 Replace cill plate

103 Reform original west cill plate corner joint connection

104 Form slip tenon to west corner post

105 Full scarf repair to the bottom of the east corner post

106 Renail studwork using similar nails to original

107 Replace long brace

108 Replace full height stud

109 Cut lower studs with a stub tenon following original detail, relocating them to the shorter position

110 Replace the missing longer stud

111 Pack under the new cill plate with lime putty mortar

112 Fix external coach screwed reinforcing plate to the arcade plate bridle joint: allow 600 × 75 × 6 mm plate: secure with 6 No. M12 × 75 mm coach screws

WEST ELEVATION

200 PLINTH WALL: As section F10

201 Carefully rake out and repoint loose pointing

202 Rebed loose and unbedded brickwork to following existing coursing

250 CONNECTION OF CILL PLATE TO NORTH AND SOUTH CILL PLATES: As section G20

251 Form slip tenon connections to north and south cill plates

252 Apply blocking pieces to north corner between the cill plate connections, prior to repegging corner joint

NORTH ELEVATION

300 PROPPING

Prop the elevation with the braced scaffold and needles

310 REPLACE DECAYED SOFTWOOD CILL LENGTH: As section G20

311 Replace decayed length of softwood cill with new oak section scarfed to original cill member (split ring connectors not required)

312 Nail studwork using similar nails to original: allow for replacing full stud length

313 Cut lower studs with stub tenon following original detail, relocating studs in shorter position and replacing missing longest stud

314 Pack under the new section of cill plate with lime mortar: as section F10

315 Fix metal strap externally to reinforce earlier wall plate repair at east end. Allow 600 × 75 × 6 mm plate: secure with 6 No. M12 × 75 mm coach screws

EAST ELEVATION

400 PROPPING

Jack up central axle support posts to soffit of tie beam (do not proceed until north and south wall scaffolding braces structure)

410 REPLACEMENT OF DEFECTIVE CILL PLATE

411 Remove defective cill plate

412 Excavate trench foundation, the depth to suit the ground conditions at 3 bricks below finished ground level

413 Repair feet of braces and posts. Reform bearing shoulders and tenons. Allow for scarfing new ends to timbers: as section G20

414 Insert new cill plate to dimensions of the original plate: as section G20

415 Build up brickwork beneath the plate using lime mortar: as section F10

420 FRENCH DRAIN

421 Lay French drain to fall at bottom of trench and direct water away from building

422 Cover drain with free-draining material with rammed chalk-finished surface below level of cill plate

G20 Carpentry repair

TIMBER TYPES

100 TIMBER FOR REPAIR

- Timber species:
  Repair of oak members: English oak: boron preservative treated if necessary as clause 150
  Replacement of structural softwood: Douglas fir: pressure impregnated as clause 160
  Repairing of softwood joinery members: Unsorted redwood: pressure impregnated as clause 160

- All timber to be free from structural defects, without spiral grain, large through-knots or rot
- Moisture content:
  Unless specified otherwise, selected timber shall be air dried to match the moisture content of the timber to be repaired
  Larger section sledge scarf repairs may be made with freshly sawn oak if air-dried material is not available

110 TIMBER FOR REPLACEMENTS

- Timber species: English oak: boron preservative treated if necessary as clause 150
- All timber to be freshly felled and free from structural defects, without spiral growth, large through-knots, ring shake, star shakes or rot
- Conversion: Freshly sawn: follow the conversion from the log dimensions of the member to be replaced
- Curved sweeps to be used with similar radius for replacement shaped members

120 WEATHER BOARDING

- Base: Existing timber frame
- Types of existing timber:
  Original oak weatherboarding to remain *in situ*
  Softwood weatherboarding to be removed
- Type of replacement timber: Air-dried oak: weatherboarding: dimensions and profile[s] as existing
- Treatment of base: Prior to fixing replacement weatherboarding: treat external surfaces of timber with proprietary insecticide-fungicide
- Fixing replacement boarding: Apply tapered timber kickers, as clause 100, where required to bottom course of weatherboards: follow present detail visible on west elevation
- Completion: Paint external surfaces of boarding with tar-based paint to match remaining original boards

130 SIZES OF TIMBER MEMBERS

- All replacement members shall be of dimensions to match those of the member to be replaced
- All new members used to replace those which have been lost shall be of equal size to adjacent similar members; where no adjacent similar member exists, the size shall be agreed with the contract administrator.

140 CONVERSION OF TIMBER

- New members shall be converted to match the figure of existing adjacent members
- This may include as appropriate:

Baulking from log with side axe

Sawing by hand, chain saw or band saw

- Circular saws may not be used for work which will be visible in the finished work

WORKMANSHIP

200 DESIGN AND FIXINGS OF ALL REPAIRS

To be agreed with the contract administrator

- All repairs to be sympathetically finished following the contours of the repaired timber
- All repairs to be dry fixed and secured with stainless steel fixings
- Stainless steel fixings should be counter bored and cross-grain pelleted

210 JOINTING

- All joints shall be executed as shown on the drawings
- Where not shown, assemble with tight, close-fitting mechanical joints such as scarves or mortise and tenon, etc.
- Butt joints reliant on adhesives or fixings for their integrity will not be acceptable

220 PEG FIXINGS

All joints between primary structural members shall be fixed with oak pegs

- Peg fixings shall be asymmetrically bored so that driving through shall draw the shoulders of the joint tightly together

230 NAIL FIXINGS

For secondary members and for fixing or refixing of weatherboarding

From 'The Cut Nail Range' supplied by the Glasgow Steel Nail Company.

240 STRUCTURAL METAL PLATES, BARS AND FIXINGS

- To be agreed with the contract administrator
- To be of stainless steel unless specified otherwise

250 WORKING OF STEEL

Exposed arises of all visible straps to be bevelled

260 FINISHING

All visible metal work to be painted black unless specified otherwise

Specification author: Peter McCurdy, McCurdy & Co. Ltd.

COMMENTS ON THE RESULT

The conservation principles were respected and feedback was very positive both during the project and after completion.

The condition survey of the timber-framed building by an independent expert was essential, particularly as external funding was to be sought. As the project was relatively small, local timber framers were employed, who had already undertaken excellent work for the Trust, and who could be trusted and relied on to adapt the specification depending on what they uncovered as the work progressed.

The level of craftsmanship was considered to be first-rate. As a general point, it is worth noting that a small contractor may find current health and safety requirements challenging which can require more time and input from Trust staff than would happen with a larger contractor. This may be offset, however, by a greater level of confidence in the workmanship and less need to monitor the work.

Chapter 4

Stone roof coverings

Because roof coverings tend to be replaced more frequently than protected fabric such as the roof frame, historical materials are less often found *in situ*. Nevertheless, the roofing specifications in this chapter all include the English Heritage guidance regarding the importance of recording existing roofs prior to stripping: the Fleece Inn specification stipulates this in general terms and the Lytes Cary one specifies numbering each slate after recording gauge, lap and margin for every course. The Belton boathouse specification requires a detailed form to be filled out for each slate removed, which may only be practical or required on very sensitive projects. All three specifications describe the handling of existing roof slates in detail: how they are to be removed, how they are to be stored and how they are to be reused. This attention to specifying the recording of historical fabric prior to removal, and the methods of removal and storage, is important since this work may not be done by conservation specialists but by general contractors.

It is almost certain that in re-roofing historically significant roofs, new or reclaimed materials will have to be introduced if existing materials cannot be repaired. The preference is always for new materials as this helps to promote traditional crafts. If this is not possible and reclaimed materials are considered, these must come with a full written provenance provided by a reputable company to avoid any risk of acquiring stolen property. Reclaimed materials are difficult to match as they are likely to have aged differently to existing material; it is better to choose new material that is authentic to the site as this should age in the same way.

All three specifications stress the importance of matching the characteristics of new slates to the existing ones. The Belton specification makes the important point that this is not simply a question of initial appearance but also the material's durability and how it will weather relative to the existing slates. The Lytes Cary project names the specific quarries to be used. The Fleece Inn and Belton specifications do not rule out the use of reclaimed material if it meets requirements but under no circumstances without a full provenance, in writing and verified.

Once again, the importance of augmenting written specifications with visual aids is exhibited, with both the Belton and the Lytes Cary specifications making regular reference to drawings.

The unusual fish-scale shape of the Belton boathouse slates required a sample mock-up panel, showing widths, lengths and laps, to be submitted for approval. Sampling is important in conservation specifications. It can be used to test methods once the contract has begun, as in the Belton case, or used at the pre-contract stage to help select a contractor and operative who can do the work to specification.

Fleece Inn, Bretforton, near Evesham, Worcestershire

CONTEXT

The Fleece Inn is a quintessential English country inn in the centre of the village, opposite the church and between the village green and a brook where it is flanked by Cotswold stone cottages. It is a Grade II-listed black-and-white half-timbered house built in the 14th century and has Cotswold stone walling along with the timber framing, with a part-thatched, part-Cotswold stone roof.

An archaeological survey has dated the original structure to around 1400. Originating from a two-bay farmhouse (one bay for human occupants and one for livestock), the building became a long house of five bays with a parlour at the north end, the hall in the middle and livestock at the southern end, where the floor is stepped down in the fifth bay. Further alterations were carried out in the 17th and 18th centuries. It was extended, with floor structures inserted and the roof raised at the southern end, culminating with the brewhouse of c.1800. It first became a licensed house in 1848.

The Fleece Inn was bequeathed to the National Trust in 1978. Since many public houses have been extensively altered and rethemed several times, the intact nature of the Fleece Inn is rare and very significant. The two principal beamed interiors, the brewhouse and the pewter room, are famous internationally. Though the brewhouse may be as late as the early 19th century, it was built in a traditional form.

The inn suffered a major fire in February 2004 which began in its thatch roof. The fire took hold rapidly and spread to the adjacent stone tile section of roof. Despite the best efforts of the fire service, it took several hours to bring the fire fully under control. The worst damage was confined to the roof, which was 80% destroyed, and the first floor, which provides domestic accommodation. The thatched single-storey later extensions on the southern end, which house the lavatories and cellarage, were also largely destroyed, with only the walls remaining.

The principal range of the roof, the southern extension and porch roofs were all covered in natural stone slate laid in diminishing courses with stone or reconstituted stone ridges. Many were destroyed by the heat of the fire or physical damage where slates needed to be ripped off to fight the fire.

On the ground floor, the three principal public rooms were only moderately damaged. Thanks to the efforts of the licensee, villagers, Trust staff and firemen, who formed a human chain, 95% of the Trust-owned contents of these rooms were saved. Plasterboard in the ground-floor ceiling from a 1960s refurbishment helped to save this floor by acting as a firebreak.

Following the fire, the Trust's Executive Committee approved plans for a major restoration and improvement of the inn. This included complete rebuilding of much of the roof and extensive repairs to the first floor. The project also included improvements to the kitchen facilities and disabled access. A Trust project team was set up to examine available

Figure 4.1 Reslating and rethatching complete. © John Goom Architects.

options and their likely impacts. Alterations to the building required Listed Building Consent, approval from the Trust's Architectural Panel and the insurance loss adjuster.

After the essential work of clearing and securing the site, a structural engineer was commissioned to carry out an initial survey of the building. Initial findings were that the building could be restored rather than rebuilt. Following this, a detailed archaeological recording was completed to assist and inform the architect in his specification and detailing of repairs. A full library of pictures, both recent and historic, was also produced and made available to the archaeologist and the architect. Photos taken during the quinquennial inspection conducted shortly before the fire were also helpful. Although the fire destroyed large areas of the roof, a bat report was still required to confirm whether or not a roost existed in the remaining roof areas.

CONSERVATION PRINCIPLES APPLIED

The Trust restoration objective was to restore the building to as near as possible its condition prior to the fire and retain as much of the original fabric as was practical, especially in the historic interiors of the public rooms. Reinstatement of fire-damaged areas was on a like-for-like basis except where some modern materials had existed that were inappropriate for historic building conservation (e.g. traditional lime plaster was used in place of the gypsum plaster that was there before).

The opportunity was taken to make some pragmatic improvements compliant with the requirements of access and environmental health regulations, however. Modern materials

Figure 4.2 Fire damage to roof structure and remnants of original slating. © John Goom Architects.

Figure 4.3 Fire damage to roof structure. © John Goom Architects.

Figure 4.4 Fire damage to roof structure. Slates removed and saved for reuse. © John Goom Architects.

Figure 4.5 New roof structure spliced in with original members. © John Goom Architects.

were introduced where it was accepted that they improved the vapour permeability of the structure, such as the installation of a roofing membrane. Any new construction was to have a minimal impact on the footprint of the existing building and was to be of simple and unimposing design.

The design approach took account of historical evidence to inform repairs and form of structure, by liaison with Trust archaeologists and curators. The project drew on a range of specialised skills and techniques in timber-frame construction. It was also seen as a way to foster traditional craft skills and building conservation techniques and provide learning opportunities.

Most of the roof, including nearly all the common rafters, required replacement and at the northern end of the building, where the damage was greatest, the fabric of walls and floors also needed repair or replacement. A fine balance was struck between the need to complete physical building work and the aim to hold on to the inn's sense of history. This meant retention of as much historic fabric as possible, even where damaged, to ensure that the building continued to tell its story. This required an informed approach, guided by good archaeology, historical research and sensitivity to the local style in all its forms.

Approximately 25% of the stone slates could be salvaged. New slates were sourced from a local quarry and were supplied sorted rather than random. Although this route is more expensive initially, it has very little wastage and the two costs end up similar. The new slate needed to be toned to blend in with existing slate. A temporary roof was extended to cover the entire building and all remaining slates were stripped and stacked for reuse. Additional platforms were added to the scaffolding to provide storage of the slates.

Figure 4.6 Reslating with areas of reused original slates and areas of new slates. © John Goom Architects.

Figure 4.7 Reslating of ridge.
© John Goom Architects.

Figure 4.8 Completed reslating with new and reused slates along with rethatching. © John Goom Architects.

Figure 4.9 Fleece Inn enclosed in scaffolding but open for business, albeit in a neighbouring barn. © John Goom Architects.

REPAIR SPECIFICATION FOR COTSWOLD STONE SLATING, THE FLEECE INN

1. Preparation

1.1 The contractor must allow for all scaffolding and access ladders to comply with all health and safety requirements.

1.2 When stripping existing roofs protection must be provided to prevent damage from falling material to any ceilings or other structure beneath.

1.3 Prior to stripping any existing slates the contractor is to record the existing course widths and particularly note the size of the largest slates.

1.4 The contractor shall strip roofs with the greatest of care in order to avoid any damage to the natural slates. All slates shall be carried from the scaffolding and no slates, whether damaged or not, shall be thrown down.

1.5 In preparing his tender and giving the estimated contract period, the contractor should ascertain and take into account likely delays in obtaining supplies of slates.

2. Materials: slates

2.1 The range of slate sizes must be set to approximately match those on the existing roof slope. If the largest existing slates cannot be reused at their full size new slates of the original size must be obtained to maintain the scale of the slating.

2.2 Special attention must be paid to matching sizes exactly where the phasing of the work necessitates abutting old and new slating on the same slope.

2.3 Damaged existing slates from on site may be redressed to form smaller slates but new larger slates must be provided to maintain the scale of the slating courses.

2.4 Sound salvaged slates from other sources to match the existing may only be used if agreed with the architect beforehand and the source of the slating is identified in writing and the source can be verified.

2.5 All new slates are to be of the best quality, properly squared, free from spots, and of such quarries and colour as shall be approved by the architect.

2.6 New slates must be selected to be of similar thickness to the original slating.

2.7 All slates must have all exposed edges (including the head if visible from beneath) hand dressed to match the existing slates.

2.8 Each slate to be holed, near the head and secured with oak, or alloy slating pegs (as specified) over the batten. Two or more pegs to be used on slates wider than 500 mm.

2.9 When alloy pegs are used new holes should be drilled to be slightly larger than the peg. For oak pegs the holes should be handpicked.

3. Materials: other

3.1 Oak pegs are to be cut oversized and tapered to fit the slate holes to avoid falling out when dry.

3.2 The underlay is to be a roofing membrane as in BS 5534 Part 1 1997 as appropriate for the type of roof construction; selection to be agreed with the Contract Administrator.

3.3 Battens generally are to be in selected softwood. The width of the batten should be at least 10 times the thickness of the nails.

3.4 Batten nails to be copper or stainless steel nails with annular rings.

3.5 Unless otherwise specified ridges are to be covered with stone ridge pieces either the existing reclaimed or new cut from a stone to match the slates as closely as possible.

3.6 Where bedding is specified the mortar shall be ungauged lime mortar as elsewhere described. Surplus mortar droppings must be carefully brushed from the slates before it has time to mark the slates and the roofs are to be left in a clean and tidy condition.

3.7 Where specified a torching mix is to be 1:3 lime and sand with addition of long animal hair at a rate of about half the volume of the lime.

4. Workmanship: slating

4.1 Slates are to be laid to diminishing courses similar to the existing but with a head lap not less than 100 mm and a minimum side lap of 75 mm.

4.2 Allow a double course of slating at eaves. This is to be set to project to the centre line of any guttering.

4.3 Verge slates are to be slightly tilted to reduce run-off.

4.4 Valleys to be formed as traditional swept valleys unless otherwise instructed. Allow for providing a suitable valley board.

4.5 Where lead-lined valleys are specified appropriate tilting fillets are to be inserted.

4.6 The drilling of old slates must be done with care and it will generally be found that a special slow-speed or hand-operated drill is the most effective for this purpose.

4.7 Where nail/peg holes have become overenlarged but the slate is otherwise sound, fix with tough plastic pegs in conjunction with combined plastic sleeve and double washer. Alternatively, reduce holes by filling with an epoxy-stone powder amalgam then carefully drilling the amalgam to form the desired hole diameter.

4.8 No half slates will be permitted as verges or eaves and the contractor is to allow for obtaining new matching or salvaged slate-and-a-half and uncloak slates for this purpose.

4.9 Cut slates are to be kept to a minimum and no slates should be cut less than half slate width.

4.10 All cut half slates – which are likely to occur at the junction of the roofs with vertical upstands – shall be nailed.

4.11 Unless otherwise specified the slating is to be left unbedded to allow free drainage except the undercloak course at eaves which is to be bedded on lime:sand mortar.

4.12 Where slates are mortar bedded they shall be soaked in water first to control suction.

5. Workmanship: underlay

5.1 Where slates are all nailed underlay is to be provided.

5.2 Underlay to be laid parallel to eaves with horizontal joints lapped over at least two rafters (75 mm if continuously supported), vertical joints not less than 150 mm.

5.3 At the eaves pull underlay taut, dress over the undercloak course and into gutters. Where specified, the underlay is to be finished on the undercloak course and dressed over a Code 5 lead eaves flashing which is to project into the gutter.

5.4 At verges, lay underlay to finish neatly slightly recessed from the outside face so as not to be visible.

5.5 At abutments lay underlay to underlap flashings by not less than 100 mm to give effective weathering.

5.6 At valleys lay 600 mm wide underlay strips to underlap general underlay.

5.7 At hips lay 600 mm wide underlay strips to overlap general underlay.

5.8 At ridges lay a length of underlay over ridge or overlap general underlay not less than 150 mm.

6. Workmanship: battens and fixings

6.1 Battens are to be set out to suit the diminishing slate courses.

6.2 Fix additional battens at ridge and eaves to allow double tiling.

6.3 A tilting fillet is to be provided at all projecting eaves. At clipped eaves the foot of the rafters must be set to allow the front edge of the wall top to act as a tilting fillet.

6.4 Unless otherwise specified, slates are to be pegged.

6.5 Where slates are held by pegs, top three courses at all ridges and every sixth course generally are to be fixed by nailing.

6.6 Where slates are pegged, torching is to be applied to the rear face unless otherwise instructed.

7. Workmanship: abutments, hips and ridges, etc.

7.1 Code 4 lead soakers are to be provided at all raking abutments.

7.2 Unless otherwise specified, allow for covering the soakers at abutments with ungauged lime mortar flaunchings to be kept as small as possible.

7.3 Unless otherwise specified, provide Code 6 lead flashings at abutments at the top of all roof slopes.

7.4 Unless otherwise specified, hips are to be mitred, provided with Code 4 lead soakers and covered with a neat mortar fillet in haired lime mortar.

7.5 Ridge pieces are to be bedded on mortar with the mortar recessed so as not to be generally visible.

7.6 Ridge pieces are not to be pointed.

8. Roof ventilation

8.1 Where insulation is included in the roof or ceiling structure, roof ventilation is to be provided by use of roofing membrane (choice to be agreed with the contract administrator) with additional ventilation and other measures as specific to the project.

Specification author: John Goom, John Goom Architects

COMMENTS ON THE RESULT

Assessing the quantity of new stone slating required cannot be done until existing slates have been stripped. A visual assessment can only give a guide. However, deliveries of stone slating can be slow and depend on both demand and availability of appropriate stone at the quarry. It is always best to build in significant time between stripping and reslating for delivery of the new slates.

The new Cotswold stone slates came from the Tinker's Barn Quarry near Naunton, Gloucestershire. This source proved to be a good match to the earlier slates which almost certainly came from a variety of sources in the North Cotswold area. The new slates weathered to a similar tone to the existing within a couple of seasons.

The quarry used is unusual in supplying stone slates already sorted into sizes. Stone slates are normally delivered as an unsorted collection sufficient to cover a square 10 ft × 10 ft (approximately 9.25 sq m). As the slater needs to set out the slating battens to suit the quantities of each slate size available, it would not normally be possible for them to assess the gauge of the battens until the new slates had been delivered and sorted. Having slates already sorted to a known size saved time.

On this project there were sufficient existing slates to cover some complete roof slopes, such as the slope of the brewhouse facing the main entrance and several of the smaller roof areas such as dormers. In view of the different life spans to be expected from old rather than new stone slates, the two were not mixed on the same roof slope.

The Boathouse, Belton House, Grantham, Lincolnshire

CONTEXT

Belton House was built in 1685–88 in the Anglo-Dutch style, with several phases of improvements between 1810 and 1840.

The boathouse was designed by the noted Gothic Revival architect Anthony Salvin, as a centrepiece of the newly extended pleasure grounds and creation of two small lakes. Whilst many landscape parks of the 18th and 19th centuries had lakes complete with boathouses, few examples remain today.

The 1821 drawings show that its main room was designed as a fishing room, with a projecting balcony and rustic twig-work balustrade. Underneath there was space for a boat. This pretty but impractical arrangement was probably not realised, as early photographs show a balcony constructed with robust flat timber balusters and a separate boat shelter alongside.

The design of the building uses all the main materials of construction in a decorative manner. The roof is of Collyweston stone tiles, each hand-shaped into a fish scale. The outer faces of the walls were pargetted with a basket-weave pattern; the joinery is all wood-grained and the leaded light windows have a decorative latticework of leads and both tinted and etched glass.

Figure 4.10 Belton boathouse. © Nick Cox Architects.

129

Figure 4.11 Belton boathouse in the late 19th century. © Nick Cox Architects.

Figure 4.12 Early sketch drawing of the boathouse by Anthony Salvin. © Nick Cox Architects.

The boathouse had occasional use in the 19th century but fell into decline during the 20th century. The balcony disappeared, the roof was covered with corrugated iron and the building also suffered vandalism.

CONSERVATION PRINCIPLES APPLIED

Prior to carrying out any works, detailed research was undertaken on site and on the archives to understand the building's history and materials. The lake level was lowered and a temporary dam built around the boathouse to enable archaeological survey and recording. A number of Salvin's drawings held in Belton's archive provided useful references to those parts of the boathouse that had disappeared, in particular the 'gallery' over the lake. A detailed analysis of the paints, plasters and renders determined the paint history of the building and nature of the materials used.

Studies were carried out, with craftsmen, to explore ways of reinstating the Collyweston fish-scale roof. A number of original tiles were in store at Belton House and these provided a pattern for restoring the roof finishes. The tiles were sufficient in number to show that while the length varied, the width was always consistent, which was necessary to enable the points of the tiles to align. The existing tiles were notably thin and consistent in their thickness. An initial effort to source new Collyweston tiles was subject to a sample area of roof being worked using the new material. These tiles were found to be neither thin enough nor consistent enough in thickness to be usable for the roof.

A second-hand supply, of known provenance, was secured and used for the work. This material was hand-dressed to make up the shortfall of roof tiles. This is a very unusual activity, possibly unique, and required great skill from the craftsman employed. The dressing of the tiles was also out of the ordinary in that the fish-scale shape had a slightly curved profile. The project benefited greatly from the skill of the roofer who had a particularly good feel and touch for working the shape. The setting out across the roof, as well as the gauging down it, were critical. All the tiles were 10 inch wide and had to fit exactly to the overall width of the roof with an appropriate projection at the verge. The pattern of the tiles also required half ties at the verge.

SCHEDULE OF WORKS AND REPAIR SPECIFICATION FOR REROOFING WITH COLLYWESTON SLATE TILES, BOATHOUSE, BELTON HOUSE

Note: Extract from Schedule of Work: sections 1–6 omitted.

Schedule of Works

7. Roof works

7.1 Remove the existing corrugated iron roof coverings and associated temporary plywood boards and cart away from site.

7.2 Allow to record the position of and label each and every roof structure element.

7.3 Allow for liaising with the architect and structural engineer on site to agree the repairs to the roof structure.

Figure 4.13 Belton boathouse before the restoration works with no gallery, corrugated roof and general dilapidation. © Nick Cox Architects.

Figure 4.14 Restoration work in progress with lake level reduced and temporary roof overhead. © Nick Cox Architects.

7.4 For the purposes of tender, allow for the scope of repairs as indicated on the drawings (as noted below).

7.5 Allow for a mock-up of the eaves as noted in the 'samples' section.

7.6 On completion of the roof structure repairs, include also for the provision of the soffit boarding as detailed in drawing no. 1034/45. This is to include a continuous, visible, tilting fillet along the full length of the eaves.

7.7 Supply and fix new roofing membrane (to be agreed with the contractor administrator) underlay, to both roof slopes. To be installed strictly in accordance with the manufacturer's recommendations, to sag between the rafters.

7.8 Supply and fix new treated softwood tiling battens to be secured with stainless steel nail fixings as the specification.

7.9 Supply and fix new Collyweston stone slate roof. The stone tiles are to be 10 inch wide and are to be laid in diminishing courses, ranging in length from 12 inch to 18 inch in 2 inch increments (subject to the check on existing tiles detailed below).

All of the tiles, except for the eaves tiles, are to be shaped to a fish-scale pattern to match the existing (see section on 'samples'). Some samples of the existing tiles will be made available for the contractor to inspect.

7.10 Allow for the supply and fix of all new Collyweston stone slates to both roof slopes. To be fixed with stainless steel nails.

7. 11 The setting out is to be extremely carefully worked out prior to fixing and as discussed with the architect. It is to be noted that the widths of the slates are all equal and therefore the spacing will need to be carefully set out to suit the size of the existing roof structure. The contractor should allow for working half slates at the verges.

7. 12 The contractor is to allow for inspecting and sorting through the tiles that are in store in the woodyard. Whilst it is not envisaged that the existing tiles will be sound for reuse, they are to be used as a point of reference for the size, shape and setting out of the new tiles. The mortar shadow lines are to be noted as a reference for the gauge and headlap of the tiles. Tables for recording the information are included in the Stone Roof Specification.

7. 13 Following sorting and laying out to assess gauge and headlap, the contractor is to report the findings to the architect. Should the headlap be considered inadequate then the slate length may need to be extended but without this altering the finished pattern for the roof.

7. 14 For the purposes of the tender it may be taken that the headlap is to be 75 mm from the point of the fish scale to the relevant tile fixing in the next course but one. For the purposes of the tender allow for the following courses:

Tile size	No. courses	Coverage	Margin
18 inch	7	35 inch	5 inch
16 inch	9	39 inch	4 1/3 inch
14 inch	9	33 inch	3 2/3 inch
12 inch	9	27 inch	3 inch

Courses and quantities to be reassessed as noted above.

7. 15 Include for the supply and fix of the square edged slates to the eaves undercourse on both elevations.

7. 16 The slates are to be bedded at their bottom edge on lime mortar but not fully pointed up to allow ventilation to the tiling battens.

7.17 Include for the provision of Code 5 lead soakers and lead flashing to the chimney.

7.18 Supply and fix a new 6 lb sand-cast lead ridge and timber roll as drawing no 1034/45.

Specification – stone-tiled roofs

1. Standards

1.1 Generally, all roofing work is to comply with the Code of Practice BS 5534; slating and tiling, subject to any qualifications below.

2. Generally

2.1 No existing slates are to leave the site without the express agreement of the employer/architect.

2.2 Repair works to existing roofs are to take into account the existing line, shape and undulation of the roof. The aim of the work is not to straighten out the existing crooked structure but to put the roof in a sound state of repair.

2.3 All work is to be finished to a sound, weather tight condition.

2.4 Bats: if evidence of the presence of bats is found then refer to the architect.

3. Investigation and recording

3.1 Where an existing roof is to be stripped and relaid, details of the existing slates, sizes, batten positions and the like are to be recorded. Tables for recording construction details and materials are appended to this specification. Copies of the completed records are to be handed to the architect following the recording work.

3.2 When undertaking recording, note should be made of whether or not slates have been bedded or torched. The position of mortar lines for bedding may also give an indication of the setting out and headlap.

4. Storage and sorting

4.1 The stripping of the roof slates is to be undertaken carefully, to avoid any damage to the existing material.

4.2 Except where a scaffold has been specifically provided for storage at high level, allowances should be made for taking the slates to ground for storage and sorting prior to refixing.

4.3 The removal of slates should commence at the ridge and during the process slates should be inspected for damage and delamination. Damaged slates are to be set on one side, for redressing where possible.

4.4 Stone slates are to be stacked on end, in piles sorted by length (measure from the fixing hole to the tail). They are not to be stacked flat.

4.5 Following sorting for length, slates should be sorted for thickness where necessary.

4.6 The sorted slates, including those considered to be damaged and set on one side, to be made available for inspection by the architect.

4.7 The contractor is to provide details of any quantities required to make up the shortfall of slates to enable the roof to be relaid.

5. Materials

5.1 Stone tiles

5.1.1 Following sorting, as detailed above, all sound material is to be identified for reuse.

5.1.2 New/replacement stone slates are to match the existing ones as closely as possible in terms of their geological type, colour, texture, size, thickness and edge dressing. Should it be necessary to source slates from a different geographical area then slates of geological and visually similar appearance are to be identified. Further, the slates are to have suitable weathering and durability characteristics as well as an ability to be finished in the manner of local tradition of the roof on which the works are being undertaken.

5.1.3 Wherever possible new slates rather than second hand are to be provided.

5.1.4 Reclaimed slates will only be considered for use if they are provided with details of their provenance.

5.1.5 Substitute materials, such as reconstituted stone, are not acceptable.

5.1.6 All undercloaks are to be of natural stone slate, as the main roofing finish, unless agreed otherwise.

5.1.7 New stone slates are to be split in the traditional manner and are not to be sawn into thickness.

5.1.8 The slates are to be dressed rather than sawn to size. Where materials are in short supply and dressing the slates, if shaped, may result in an unacceptably high level of wastage then sawing first may be considered acceptable providing the stone tiles are subsequently hand dressed to finish the edges.

5.1.9 Where existing sound material is to be reused, it is to be reused in its original orientation. Do not turn slates.

5.2 Nails

Nails are to be to BS 1202. Nails fixing battens and stone roof tiles are to be stainless steel, of adequate length and gauge to provide a secure fixing. The fixings are to be made without splitting the tiling battens.

5.3 Tiling battens

5.3.1 Tiling battens are to be of treated softwood. CCA treatment is not to be used in residential or domestic constructions. The battens are to be free from decay or insect attack, free from any splits and shakes; wane is permissible on one arris only; knots or knot holes are permissible if less than one-third of the width of any sound surface; the moisture content is to be below 20% at the time of fixing.

5.3.2 The battens to be pressure treated (vac-vac). Treatment is to be carried out using only wood preservatives approved under the Control of Pesticides regulations and by a member of the Wood Protection Association.

5.3.3 The treatment is to be clear, without dyes added.

5.3.4 Batten sizes are to be 50 mm × 25 mm.

5.4 Underlay

The underlay is to be agreed by the contract administrator. The roof underlay is to be installed strictly in accordance with the manufacturer's recommendations.

5.5 Mortar

Where mortar is required for bedding ridges or for back bedding or spot bedding tiles then it is generally to be based on a 1:3 eminently hydraulic lime:sand mix. Samples of the mix are to be prepared for inspection by the architect. Refer also to the specification for mortar.

6. Workmanship

6.1 Stripping, sizing and stacking: work is to be undertaken as detailed in the section above.

6.2 Care is to be taken when stripping and relaying the roof to ensure that there is no eccentric loading imposed on the roof structure.

6.3 Protection

Where a roof is to be stripped, provide all necessary protection, boarding, tarpaulins, dust sheets and the like to ensure that no damage occurs to the property or the contents of the property. Such damage is to be made good without cost to the employer. It is particularly important to provide protection within the roof space to catch any debris, particularly torching if present, prior to stripping the roof.

6.4 Cleaning

Following stripping of the roof, properly clear the work area of all debris created during the course of the stripping operation. Include for denailing the rafters as necessary to enable refixing.

6.5 Inspection

Following removal of the roof finishes, allow for inspection of the roof structure with the architect and report any defects encountered during the roof stripping process.

6.6 Underlay

Supply and fix the underlay in accordance with the manufacturer's recommendations. Include for laps and taped joints as detailed in the manufacturer's recommendations.

6.6.1 Roofing membrane, agreed by the contract administrator, is to be laid in accordance with the requirements of the BBA certificate for the product.

6.6.2 Where the underlay is directly below the tiling battens, it must be allowed to sag slightly between the rafters to allow for drainage.

6.6.3 The underlay is to lap over the top of the ridges by at least 150 mm down each side of the slope so as to produce a 300 mm wide double layer along the centre line of the ridge.

6.7 Battens

Supply and fix new battens, as specified, to be secured with stainless steel nails with minimum nail penetration of the rafters of 50 mm. The battens are to be fixed at centres to give the correct lap of the stone roof tiles. The ends are to be cut square and centred over the rafters and nailed without splitting. No more than 1 in 3 batten joints should occur on any one rafter.

6.7.1 Where necessary, allow to pack the battens (MDF, hardboard and the like are not acceptable) to maintain their level/contact with the rafters. Use longer nails where necessary to ensure the minimum penetration specified above is achieved. If the packing is greater than the thickness of a batten then refer to the architect.

6.7.2 Provide additional battens, sprockets, tilting fillets at the eaves, ridge and other perimeter features as necessary.

6.7.3 Battens are to be not less than 1.2 m in length and laid over a minimum of 3 No. rafters.

6.8 Slating gauge

6.8.1 The contractor is to assess and confirm to the architect the number of courses of each slate length that can be achieved with the salvaged and new slates. Where all new slates are specified then the number of courses are to be as detailed in the Schedule of Work.

6.8.2 Particular care is to be taken over assessment of the width of the slates, particularly where a standard width is in use for the roof, to ensure that the overall length of a course fits the roof to be relaid.

6.8.3 The maximum/minimum slate sizes are to be as detailed in the Schedule of Work and slates in between are to be evenly graded in diminishing courses to match the character of the existing roof work.

6.9 Headlap

The headlap is to be related to the headlap of the existing roof, where measurement of this is available. Generally, the headlap is to be a minimum of 75 mm. Refer to the architect if the existing is less than this figure. The minimum lap is to be maintained as the slate length changes up the roof.

6.10 Sidelap

The sidelap is generally to be not less than 75 mm. Where slates are shaped, the sidelap may be defined by the shape of the slates, generally half the slate width.

6.11 Slate laying

The slate should be laid one course and never more than a few courses at a time, working from thicker to thinner slates across the roof if that is the character of the roofing material in hand.

6.11.1 The slates are to be laid to avoid rocking or uneven surfaces and to avoid variations in thickness from one slate to the next.

6.11.2 When specified, bedding or pointing the tails of the slates is to be undertaken as the work progresses, by using the minimum mortar required.

6.11.3 Where underlay is used then the bedding should be provided so as to support the edges and centre of the slate with the remainder left open to allow airflow beneath the slates. Provide shales/gallets if there is a large gap beneath the slate to ensure proper support.

6.11.4 The mortar is to be finished back from the edge of the tiles and is on no account to be taken over the top edge of the slate. It should be finished approximately ½ inch to 1 inch back from the tail edge of the slate.

6.11.5 Where no underlay is used, then the slates may be bedded with the bottom edges pointed and struck back.

6.12 Backers

Backers are only to be used where slates of sufficient width lie below to provide the sidelaps as specified.

6.13 Wide slates

Wide slates should be reserved to close the half bond at verges, hips and abutments.

6.14 Underslating

Unless specified otherwise, underslates at the eaves are to be the same material as the main roof finish.

6.15 Verge slating

Allow for pointing the exposed edges of verge slating with mortar. The mortar generally should be based on a 1:3 eminently hydraulic lime:sand mix. Where possible, the verge slating should be slightly tilted to help prevent rainwater run-off from the verge. The mortar to the verges should be laid as the work proceeds to ensure a proper fill, rather than just surface pointing on completion of the works.

6.16 Nailing

Generally slates are to be twice nailed.

6.17 Stretching

Stretching of slates by opening up the width of joint between the slates, to reduce the number of courses or overall number of slates required in the length of a new course, is not acceptable.

6.18 Ridges

To be particularly specified for the project in hand. Stone ridges are to be back bedded in mortar and solid bedded at the joints. The manner of pointing is to be so as to avoid the creation of a large mortar fillet on the face.

6.19 Hips

To be as particularly specified for the project in hand.

6.20 Valleys

To be laced/swept/secret as particularly specified for the project in hand.

6.21 Gauge adjustment

Where existing ridges and eaves do not run parallel the gauge is to be adjusted to allow the courses to run the full length of the slope. Adjustments made to the gauge established are to be made evenly over the full height of the slope and not in any one particular course. In no situation is the gauge to be less than the headlap specified.

6.22 Weather tightness

The finished result of the laid roof is to be weathertight. To this end it is important for the stone slates to be properly sorted, selected and laid.

7. Completion

7.1 All internal roof spaces to be clear of all debris created during the works and are to be left in a clean state, to the satisfaction of the architect.

7.2 All rainwater goods, gutters and the like to be thoroughly and properly cleared of debris, including any gullies into which the rainwater could discharge.

7.3 The ground beneath the scaffolding is to be cleared of any debris and the like, to the satisfaction of the architect.

7.4 Any broken or damaged slates are to be replaced by the contractor, without cost to the employer, to leave the roof in perfect order on completion of the works.

Belton House: checklist for boathouse roof repairs

Stone slate roof: checklist for recording materials		
Ridges	Material	
	Dimensions	
	Shape	
Copings	Material	
	Dimensions	
	Shape	
Slates	Limestone	
	Sandstone/gritstone	
	Composition	
	Dimensions: length range/random	
	Thickness: average or range	
	Gallet or shadow material	
Battens/laths	Split	
	Sawn	
	Species	
	Dimensions	
	Fixings	
Slate fixings	Pegs: sawn or split	
	Wood: species	
	Other: bone, etc.	
	Dimensions	
	Nails: hand made	
	Nails: machine made	
	Wire	
	Dimensions	
	Materials or composition	
Mortar	Bedding	
	Pointing	
Flashing/soakers	Material	
	Dimensions	
Torching	Composition	
Slating felt	Composition	

Stone slate roof: checklist for recording construction details		
Ridges	Fixing: bedding or nailed, etc.	
Pitch		
Slating	Headlap	
	Number of courses	
	Margin for each course	
	Batten/lath gauges	
	Side lap: minimum or range	
	Gallets	
	Single lap	
	Shadow slates	
Dormers	Pitched: mono or duo	
	Hipped	
	Cheeks: slated or other	
	Junction weathering	
Abutments	Weathering: mortar or metal	
Gables	Plain or coping	
	Fixing for copings	
	Flush, overhang or oversail	
Hips	Mitred	
	Covered: stone, tile, lead, etc.	
	Detail: roll top, etc.	
Valleys	Open	
	Mitred	
	Chevron	
	Swept	
	Laced	
	Other	
Eaves	Plain or sprocket	
	Slating detail	
Rafter centres	Range or average	
Fixing system	Battens or laths	
	Slates fixed to rafters	
	Pegs or nails	
Torching	Full, partial or none	
Slating felt	Laps: head (up slope)	
	Laps: sided (across slope)	

Specification author: Nick Cox, Nick Cox Architects

COMMENTS ON THE RESULT

The roof finish looks remarkable; it was a very unusual use of Collyweston stone roof tiles. Trials and mock-ups provide an opportunity for craftsmen to develop the particular technique required for a project; they are not just for the client and architect to approve but are a means of learning. Material sourcing was an issue; development of a new source of Collywestons would be good.

In view of the fragility of the ceiling, the slates were secured with stainless steel screws rather than nails. This may prove difficult to deal with for any isolated replacements in the future. This was a matter discussed at the time and the balance of interest fell in favour of the screw fixing.

The render was also something of a 'one-off'. Initial study done off site before the project began was invaluable in assessing the technique for forming it – two coats of plaster with the top one imprinted by a mould – and enabling preparation of a suitable specification. In practice, pressing the mould onto the face of the plaster displaced too much material, therefore the mould was first filled and then pressed onto the base coat.

Areas of historic render that were only slightly friable and considered worthy of retention were kept, but these have decayed more readily than expected. The roof design does not have a gutter and splash-back off the gravel path marks the render. An alternative ground finish would be good.

Figure 4.15 The finished roof.
© Nick Cox Architects.

Figure 4.16 Trial sample of Collyweston fish-scale roof tiles. © Nick Cox Architects.

Figure 4.17 Establishing the gauge of the tiles. © Nick Cox Architects.

Figure 4.18 Template used to size and shape the fish-scale tiles. © Nick Cox Architects.

Figure 4.19 New roof being laid. © Nick Cox Architects.

The structural works were quite comprehensive and done as proper joinery repairs with spliced joints. Some resin was used but kept to a minimum. A difficult detail was the render over the timber frame running down to the stone plinth and the desire to keep the bottom part of the frame dry. A building paper was used behind the render to achieve this. As is often the case with historic buildings, what one is dealing with doesn't necessarily conform to any standard or current best practice. Discussion is always helpful to tease out a solution. No one person has the monopoly on good ideas and conversation between client, craftsman and designers can all help.

Small areas of remaining wood graining on the timber frame were uncovered and skilfully replicated by the decorator. This finish will have to be maintained by a specialist decorator. The painted timber balustrade over the water and where the finish runs down the posts into the water will probably require reasonably regular maintenance and clever access arrangements.

Small fragments of the decorative leaded lights informed the design and type of replacement glass. Some have since been damaged (vandalised), but as the building is at a distance from the main house, the glass is always going to be at risk.

Condensation and mould growth on the ceiling have been an issue owing to it being an unheated, often closed-up space over a lake and its built fabric comprises components of different thermal performance. Some additional ventilation might be useful.

Figure 4.20 Decayed roof timbers.
© Nick Cox Architects.

Figure 4.21 Roof timbers repaired by splicing in new material. © Nick Cox Architects.

Lytes Cary, near Somerton, Somerset

CONTEXT

Lytes Cary is a Grade I 15th-century manor house. The house, with its 14th-century chapel and 15th-century Tudor great hall, was much added to in the 16th century.

Neglected after the departure of the Lyte family in about 1748, the house declined during the 18th and 19th centuries. It was rescued from dereliction in the early 20th century by Sir Walter Jenner who refurnished the interiors in period style. In order to leave the historic buildings intact, he built the West Wing for his family to live in. He bequeathed Lytes Cary to the National Trust in 1949.

Hamstone tiles were used on the roof of the Edwardian extension. The source is unknown but most certainly reclaimed. A survey of the roof in October 2000 revealed signs of slight water penetration and sections of poorly executed repairs. It recommended the stripping and recovering of the roof slope, which was completed in 2004.

CONSERVATION PRINCIPLES APPLIED

There was a pressing need at a number of South Somerset properties for Hamstone roof tiles. The Society for the Protection of Ancient Buildings had notes to show that Hamstone

Figure 4.22 Lytes Cary Manor. © National Trust Images/Nick Meers.

147

Figure 4.23 Hamstone roof at Lytes Cary. © Philip Hughes Associates.

Figure 4.24 Dressing new valley tiles; old weathered tiles have been relaid on the roof slope to the left of the picture, the new tiles are on the right. © Philip Hughes Associates.

tile production had died out in the 19th century, possibly because the traditional tile beds at the quarry had been worked out. As the east slope of the West Wing pitched roof and an adjoining flat lead roof cannot be seen from the ground, Lytes Cary was considered one of the few places that a trial of new Hamstone tiles could be carried out.

Figure 4.25 The finished Hamstone roof; note bat access tile to the right of the picture. © Philip Hughes Associates.

It was hoped that the top and bottom of the blocks from the beds at the Ham Hill Quarry, then a waste product, could be used. Sheffield Hallam University carried out accelerated weathering trials on tiles produced this way, but they started to fail at an early stage. In consultation with the quarry, it was decided to cut through the block and then tool tiles to produce a similar tile to the historic tile samples. This time the weathering trials were successful with a 50 years-plus life achieved.

REPAIR SPECIFICATION AND SCHEDULE OF WORKS FOR REROOFING WITH HAMSTONE SLATES, LYTES CARY MANOR

Notes: Relevant extracts included only.

Specification of materials and workmanship

ROOFING – STONE SLATES

MATERIALS

Stone Slates   Existing stone slates are to be salvaged and either reused or set aside for reuse elsewhere by the employer as scheduled. Where stone slates are to be reused any broken or decayed slates are to be trimmed down to produce smaller slates where possible. All edges of cut down stone slates to be hammer dressed on site by the contractor. <u>New</u> stone slates to make up numbers or for whole slopes (as scheduled) are to be supplied by the employer from one of the following quarries as scheduled:

Ham Stone Slates
The Ham Hill Quarries, Ham Hill Masonry Works, Stoke-sub-Hamdon, Somerset TA14 6RW

Cotswold Stone Quarries
Brockhill Quarry, Naunton, Cheltenham, Gloucestershire GL54 3BA

Lead Slates	Lead slates are to be of Code 7 lead, sized to ensure adequate laps.
Pegs	Copper pegs for stone slates are to be 4 gauge large head copper pin of length suitable for their purpose.
	Oak pegs are to be of size suitable for existing stone slates. Oak pegs may be obtained from Carpenter Oak and Woodland, Hall Farm, Thickwood Lane, Colerne, Chippenham, Wilts SN14 8BE.
Nails	Nails are to be of copper or stainless steel, of length and weight suitable for their purpose.
Screws	Screws are to be of stainless steel, of length and size suitable for their purpose.
Battens	Battens to be of riven oak heartwood approx. 3/8 inch × 1inch selected for straightness.
Counter Battens	Counter battens to be of sawn softwood 75 mm × 25 mm treated with an appropriate preservative.
Tingles	Tingles for fixing lead slates are to be of Code 7 lead 35 mm wide. Tingles for fixing stone slates are to be of austenitic stainless steel.

WORKMANSHIP

Stripping Stone

Slates	The stone slates are to be stripped off each roof slope in courses, numbered and neatly stacked in separate piles for each course. The existing gauge, lap and margin are to be carefully recorded for every course and a copy of these notes is to be provided for the contract administrator.
Nailing	Do not nail through any plumber's metal work.

Generally

Lay stone slates as follows:

a) Patch bed stone slates to level up in mortar mix F (see pointing section), but keep mortar at least 50 mm away from bottom edge of stone tile.

b) Twice peg every stone slate over 250 mm wide, otherwise single peg.

c) Where peg holes in stone slates occur over rafter positions, fix the stone slates to rafters with nails, so that the nails will not pull through, lever out, or bend under the weight.

To roof slopes:

d) Slightly open jointed with horizontal courses following general undulation in the roof and with vertical joints centred (as closely as possible) over the stone slate below.

e) Head lap (over fixing position) as set out below:

Effective length of stone tile (peg hole to tail)	Visible length of stone tile (margin)	Minimum lap
7 inch (175)	60 mm	55 mm
9 inch (225)	78 mm	70 mm
11 inch (280)	100 mm	80 mm
13 inch (330)	120 mm	90 mm
15 inch (380)	140 mm	100 mm
Above 15 inch	(Effective length – lap, then divide by 2)	100 mm

f) Side lap minimum 150 mm (or to fall within the central third of the tile for narrow tiles).

g) Vertical joints in top course to be positioned to avoid occurring within 75 mm of vertical joints in the ridge.

h) Maintain the minimum lap for the size of stone tile at the 'twist' (change from one size of stone tile course down to the next).

i) Where stone slates are heavily 'shouldered' maintain the lap by introducing lead slates.

Abutments   Maintain side lap and avoid use of narrow stone slates. Dress in lead soakers to receive cover flashing/fillet.

a) Refix slipped or broken stone tiles as above where access is available.

b) Where an individual stone tile has slipped refix with a stainless steel tingle. The tingle to be secured with a stainless steel screw (to limit vibration and disturbance) and is to be turned over the bottom edge of the stone tile.

c) Mortar bed slipped stones tiles in mortar mix C (see pointing section) as above.

Patch Repairs

d) Patch repair broken stone tiles using a lead slate slipped in beneath the stone tile and secured with a lead tingle.

e) Patch repair areas where the head lap is insufficient by slipping in sections of natural slate and bedding in mortar mix C or using a lead slate secured with a lead tingle.

f) Patch repair areas where the side lap is insufficient using a lead slate secured with a lead tingle.

Schedule of Works

    1. Stripping of roof coverings

1.1 Strip all stone tiles, valley tiles, tile battens and underlay to roof slopes C2, C3, C4, C5 and C6 complete. Include to sort and store all reusable stone tiles on site for reuse (including damaged stone tiles which may be redressed) and cart away all battens, underlay and debris.

1.2 Allow to strip the laced valleys between roof slopes C2 and D2.

1.3 Strip all stone tiles, valley tiles, tile battens and underlay to roof slope B2 from the eastern parapet to the ridge line of roof slopes C1/C2. Include to sort and store all reusable stone tiles on site for reuse (as above) and cart away all battens, underlay and debris.

1.4 *Extra-over items 1.1, 1.2, 1.3, 1.6*
Allow to carry all removed stone tiles and valley tiles from the roof slopes to the northern courtyard and stack on pallets for removal. Retain enough stone tiles on site to retile the southern end of roof slope C2 to the ridge line of D1/D2 and reform the laced valley between roof slopes C2 and D2.

1.5 Strip the bottom 5 No. courses of stone tiles from the junction of roof slope D1 at eaves level and A2 including the back guttering to chimney CH4 to allow new leadwork to be fixed. Include to sort and store all reusable stone tiles on site for reuse (as above) and cart away all battens, underlay and debris.

1.6 *Extra-over items 1.4*
Strip all stone tiles, valley tiles, tile battens and underlay from roof slope D1. Include to sort and store all reusable stone tiles on site for reuse (as above) and cart away all battens, underlay and debris.

1.7 Carefully remove stone ridge tiles from the full length of the roof slopes B1/B2, C1/C2, C3/C4 and C5/C6 and D1/D2. Allow inspection of tiles with contract administrator. Clean off mortar and store on site for reuse. Cart away all debris.

1.8 Cut out and renew patches of decayed timber boarding to roof slopes B2,. C2 and D1 with new 120 × 20 mm treated softwood boards fixed with stainless steel nails. Assume 20 No. areas 3 boards wide and 1.5 m long.

1.9 Lift all sheet leadwork and wood rolls from the flat roof slopes A1, A2 and A3 and cart away. Include to offset salvage costs of the lead against the cost of the works.

1.10 Remove the lead roof and cheek coverings from the dormer window on roof slope B2 and cart away. Include to offset the salvage costs of the lead against the cost of the works.

1.11 Remove the lead back gutter to chimney CH4 on roof slope B1 and cart away. Include to offset the salvage costs of the lead against the cost of the works.

2. New roof covering – pitched roofs

2.1 A lightning conductor system is to be installed under the new roof coverings by a specialist contractor – see section 8. Allow time to liaise with and attend on lightning conductor installation contractors.

2.2 Fix with oak pegs, stone tiles in diminishing courses from stocks on site, on oak battens over roofing membrane (to be agreed with contract administrator) to roof slopes B2, C2, C3, C4, C5 and C6. Form a double course at the top and bottom eaves and ensure that the head and side laps are adequately maintained through-out. Make up shortages with new stone tiles supplied by the employer. (NOTE: extent of new stone tiles required is subject to confirmation – see item 2.8 below.) Assume 20%.

2.3 Immediately upon delivery include to cart the new stone tiles required in item 2.2 from the north courtyard into the central courtyard and store on timber battens prior to use.

2.4 Fix 5 No. courses of stone tiles from stocks on site at the junction of roof slope D1 and flat roof A2 fixed on oak battens over bituminous roofing underlay forming a double course at the junction with the lead. Include to make good any disturbed surrounding stone tiling.

2.5 *Extra-over item 2.4*
Fix stone tiles from stocks on site to the whole roof slope D1 in diminishing courses on oak battens over roofing membrane (to be agreed with contract administrator). Form a double course at the top, the eaves and with the junction of the lead.

2.6 Form laced valleys with stone tiles from stocks on site with a 400 × 300 mm Code 4 lead soaker at valley junctions as follows:

a) Between roof slopes B2/C2
b) Between roof slopes C2/C3
c) Between roof slopes C2/C4
d) Between roof slopes C2/C5
e) Between roof slopes C2/C6
f) Between roof slopes C2/D1
g) Between roof slopes C2/D2

2.7 Dress stone tiles to ensure adequate laps and fix Code 4 soakers at abutments as follows:

a) To the east parapet and chimney CH7 on roof slope B2
b) To the east and west cheeks of the dormer window in roof slope B2
c) To the east and west elevations of chimney CH6 on roof slope B2
d) To the west elevation of chimney CH5 on roof slopes C3 and C4
e) To the west elevation of chimney CH4 on roof slopes C5 and C6
f) To the south parapet on roof slope C2
g) To the east elevation of CH4 on roof slope D1

2.8 *Extra-over items 2.2, 2.4, 2.5, 2.6 and 2.7*

Undertake all retiling of roof slopes B2, C2, C3, C4, C5, C6 and D1 as described substituting stone tiles from stocks on site for 100% new stone tiles supplied by the employer. The south end of roof slope C2 from the southern parapet to the ridge line of D1/D2 and the valley between C2 and D2 are to be completed in stone tiles from stocks on site. (NOTE: extent of new stone tiles required is subject to confirmation – see item 2.2 above).

2.9 *Extra -over item 2.3*

Immediately upon delivery include to cart the new stone tiles required in item 2.8 from north courtyard into the central courtyard and store upon timber battens prior to use.

2.9 a) Bed stone ridge tiles from stocks on site in hydraulic lime mortar type C to the full length of ridge of roof slopes B1/B2, C1/C2, C3/C4, C5/C6 and D1/D2.

   b) Make up shortages with new Ham Hill stone ridge tiles to match – assume 15%.

2.10 Fix clay plain tiles from stocks on site with stainless steel nails into oak battens over bituminous underlay to the bottom 10 No. courses of roof slope I2 to the length of the junction with flat roof A3. Form a double course of tiles at the bottom eaves with the junction of the lead. Cut full tiles to half tiles to maintain half bonds and place new lead soakers at the abutment to the parapet wall and make good any disturbed surrounding tiles. Make up shortages with new clay plain tiles to match. Assume 15% of the tiles removed.

Specification author: Sam Wheeler, Philip Hughes Associates.

COMMENTS ON THE RESULT

The new tiles are performing extremely well so far and that bodes well for the life that will be achieved. However, the new tiles, even with the tooling, produce a much flatter roof that does not have the character of the older tiles.

Roofing operatives highly experienced with stone tiles should be used for Hamstone roofs as quality control on site is critical due to the nature of the stone. The cost was double that compared to Cotswold stone tiles owing to the high level of labour required in the production.

The quarry felt that the accelerated weathering trials were overly dramatic and based on concrete tile testing methods. It could be that an *in situ* trial of the top and bottom of bed tile could be carried out on a smaller building, such as the Priory Dovecote at nearby Stoke-sub-Hamdon.

The inserted lead bat access tiles and the insulation were also successful.

Chapter 5
Thatch

Thatch was a common form of roof covering throughout much of Britain from the Iron Age until the end of the medieval period. In many rural areas, it remained the most practical and available material until the mid-19th century. It is one of the oldest building methods still practised today. Contrary to common belief, thatch is not just a material of historic significance, it is still used in new build. The English House Condition Survey for 2005–6 records almost 35,000 dwellings roofed in some form of thatch. Most examples in England and Wales are of water reed or special varieties of straw grown for the purpose.

As one of the oldest crafts in Britain, thatching, more than most trades, has been continued with a minimum of formal specification. Instead, thatching has relied mainly on the experience and reputation of the thatcher. There can seem little point in a practitioner writing a specification to describe a process that the craftsman would have to explain to the practitioner.

The first case study, Berg Cottage, is an example of how brief the specification can be if a specialist thatching consultant is employed to manage all aspects of the project from inception to completion. The consultant's duties involved carrying out condition assessment and surveys, working alongside a specialist thatch archaeologist as part of the pre-tender site investigations, interpretation of historic information and research, sourcing of materials and approvals, contractor selection, specification, tender, review and on-site management. The consultant conducted tender visits to explain the works, discuss the approach and agree tender detailing based on the research information presented.

The consultant was commissioned to manage and check every aspect of the material sourced, practices on site, programming and managing the defects inspection recording and snagging.

The rethatching formed part of an overall larger project of conservation and repairs to the property, the roofing element being a subcontract to a main contractor. The main contractor was accountable for programming of the works, timber repairs, repairs to chimneys and structural repairs to the perimeter of the building.

The specification points out an important consideration in any rethatching project: any change of materials, finish or detailing to the existing thatch and the underlying roof structure of a listed building will require Listed Building Consent.

Although desirable, the contractor cannot always be known. The chapter continues with the second case study – Selworthy Cottage on the Holnicote Estate – which was an extensive thatching project and one of the Trust's first attempts at establishing guidelines for thatching specifications. Its primary objective was the repair and rethatching of the leaking roofs, but it also enabled a greater appreciation to be developed of historic thatching traditions in West Somerset and provided the encouragement for the estate to become self-sufficient in thatching materials and labour. It is based on an exercise by the Trust to formalise the procedure for a number of reasons, including situations where the thatching contract must be put out to tender.

The thatching work at Holnicote provided the basis on which to write a specification for thatching in combed wheat reed in the West Somerset style. This specification is not intended to over-ride or inhibit thatchers carrying out their craft in their own unique style, but it lays down the basic approach that should be adopted.

When using thatch, particularly on historic buildings, it is important to maintain regional characteristics, techniques and materials. Local planning authorities should be consulted before work on thatched roofs is undertaken, particularly where stripping of the old thatch is proposed or where the covering style is to be changed. For instance, decorative ridges have become popular in regions where they are not historically accurate. Water reed roofs tend to have a longer life-span than other thatches and this has led to problems with it being specified in locations where it has not been used historically. Its use has become a contro-versial matter and many local authorities will want to see the type of thatch traditional to their area maintained.

All thatching projects to historic buildings should begin with recording, including an archaeological investigation and inspection, as well as a general inspection by a qualified person to assess the current condition of the thatch. The investigation takes the form of documentary research, an inspection and recording of the accessible roof voids and the formation of a step trench of the roof covering to expose the layers of thatch (this latter action will not be necessary if the previous history of the thatch is known). Thatching-related debris that has fallen onto ceilings below can also provide valuable information. Alongside condition, what is sought is evidence of wheat varieties used, type of fixing materials and smoke-blackened material (thatch or rafters) which would help determine the construction and age of the roof structure (historic smoke-blackened thatch is covered by statutory protection in listed buildings).

The archaeological investigation may also involve a watching brief as rethatching takes place, as specified on the Holnicote project. It was a requirement for an archaeologist to be present and, specifically, alerted to the discovery of any smoke-blackened material. The rethatching project should always allow for retaining material from original layers possibly containing historical evidence. The Holnicote project further specified that a photographic record be made before and during the thatching to record materials and methods as base layers were exposed.

In addition, it is important to establish the presence of rare mosses or lichens prior to any works, major or minor, so that suitable steps may be taken to preserve these *in situ*.

Berg Cottage, Barkway, Hertfordshire

CONTEXT

Berg Cottage is a Grade II-listed, timber-framed, thatched cottage. It was built in the mid-16th century and extended and converted to at least two dwellings in the 17th century. It is a prominent two-storey building within the village because of its size, age and thatched roof.

The basic house is a medieval open hall, complete with wooden screens, a central fireplace and what appears to be an inserted floor over the central hall, but the structure has obviously been heavily altered both at ground level and within the upper storeys. The date stone suggests the building was built (or more likely rebuilt) in 1687. The Trust has conducted a Historic Building and Structure Survey of the property, which largely supports the view that Berg Cottage once functioned as an open hall and was significantly altered. Before it was given to the Trust in 1938, the building was called 'The White Hall' which may be a reflection of the original high status of the property. (It was renamed after the donor, Miss Madeline Berg.)

The building's main roof has a double pitch extending down to low eaves on the rear slope with two windows halfway down this elevation. The thatch on the front face descends to first-floor ceiling height but on the rear pitch descends over a rear extension to ground-floor

Figure 5.1 Berg Cottage before rethatching. © National Trust.

level. In addition, there are lower double-pitched extension roofs to both gable ends. The roof timbers are of mixed types and sizes but are predominantly common rafters pegged at the apex and supported on purlins at the midpoint.

In 1939, in one of the earliest examples in England of conservation work on a vernacular structure, Berg Cottage was restored by architects and craftsmen from the Society for the Protection of Ancient Buildings. It thus provides a valuable record of the approach and techniques that were used by conservationists at the time. During this phase of work, the building was rethatched in traditional long straw, probably produced from the winter wheat variety Maris Huntsman or Maris Widgeon. The roof was finished with a block-cut wrap-over type ridge, its leading edge cut to a detailed pattern, and pinnacles on all gable ends, which are traditional in this area.

As with all long straw, the mature wheat was cut with a reaper binder and run through the drum of a traditional threshing machine to remove the grain without crushing the straw unduly. This straw was then built up into a bed, wetted and drawn to produce cleaned 'yealms' of straw that were 'sparred' onto the existing base coat in courses using twisted hazel 'spars'.

Eventually, the apexes of the ridges were at the end of their serviceable life, as were the aprons on both sides of the chimney. The thatch was being kept in place by a cover of 19 mm 20 gauge galvanised wire netting that was in good condition. A condition survey of the roof was conducted in June 2001.

Figure 5.2 Construction of traditional 'wrap-over' ridge. © National Trust.

Figure 5.3 Exterior hazel spars and liggers at eaves and gables. © National Trust.

Figure 5.4 Exterior hazel spars and liggers at eaves and gables. © National Trust.

Only one phase of thatching was visible from ground level: an underlying 'brow' course of straw and the existing eaves course. These showed that the eaves and barges were stripped back to the underlying timber at the last rethatching. As expected, the weathering coat on the rear (south) pitch had decayed more quickly than the front (north) coat due to more active fungal decay on the southern face. Nevertheless, this top coat had performed below average in terms of longevity for long-straw roofs in this region. External hazel fixings ('liggers') just above the eaves and inside the barges had also largely degraded.

The building had a topcoat of long straw to all areas, with only one base layer of long straw. The front elevation had a double thickness of eaves and gables that were installed at the time of the last rethatch. The main coat work to the front elevation appeared to be wearing evenly, but close inspection showed that the straw was very decayed on the outer surfaces and light pressure caused the straw to collapse. The thatch on the rear elevation had a number of visible gullies along with extensive decay to most areas and overall was in a poor condition. All verges had been cut too close to the gable timbers and did not provide adequate protection to the end walls.

Accessing the roof space allowed a closer examination of the roof timbers and the interior base coat of thatch. An internal inspection showed that the base coat appeared to be in good condition, but it was not possible to ascertain its thickness and suitability for over-sparing until the outer layer of decayed thatch had been stripped. The roof timbers were mostly in a sound condition, as were most of the softwood battens (replaced in 1939).

Figure 5.5 New thatch layer being laid over existing base layer. © National Trust.

Any ancient thatch that survived in 1939 must have been stripped and replaced with a new base coat that was tied to the timberwork with lightly tarred (hempen) twine. Tarred twine was used to tie base coats of thatch to battens and rafters from *c.*1800 into the 1960s and is still used to tie in eave 'brow' courses and barges. Steel 'crooks' may well have been used to fix the thatch directly to the timbers in some locations, but none were found during the inspection.

CONSERVATION PRINCIPLES APPLIED

The relatively brief inspection confirmed that no truly historic thatch survived. The existing base coat was interesting but more important, in terms of thatching history, was the debris found in the roof void. This material provided direct evidence of previous episodes of rethatching and allowed comparisons to be made with the evidence that survives in the adjacent cottage which has the only intact ancient thatched roof in Barkway.

The old thatch, fixings and battens found as debris in the roof void could obviously only go as far back as 1687 when the first-floor ceiling was inserted. There were sections of *c.*2.5 cm round wood (occasionally with short nails) and split oak laths (*c.*5 cm wide × 1 cm, also with corroded nails) which once acted as battens, neither of which were blackened. Thus, two phases of historic battening survived: the early, post-1687 phase, which relied on high-quality split oak laths, and a later phase – or perhaps a repair – using relatively thin round wood. The old thatch was entirely threshed straw of bread wheat *Triticum aestivum*, which is in keeping with the historic long-straw and wheat-growing tradition of the region. Degraded tips of both split and round wood (hazel) spars also survived, which suggested that the roof was regularly recoated with fresh straw as in the present day. The first post-1687 base coat was tied to the battens and rafters using thin, twisted 'withes' of hazel or willow, as several knotted bonds – decayed only where they met the weathered surface – also occurred within this heap of stripped thatch.

The existing base coat was significant only in that it provided an example of thatching methods and type of long-straw thatch in use in Hertfordshire in 1939. Listed building consent required a correct thatching style and material pertinent to the area, but because only a relatively modern base coat survived, it was readily determined that any 'point' repair work would not disturb thatch or fixings of historic significance.

The outer existing thatch base coat and its fixings were dry, solid and apparently thick enough to provide a good foundation for a new weathering coat of good-quality long straw using a traditional spar-coating technique.

Given the delicate condition of the topcoat of thatch on the front elevation and the level of decay on the rear slope, simply re-ridging was not an advisable option. In thatching terms, the thatch was not particularly old and would have been expected to last a little longer than it had. A possible cause for its shortened life may have been an overuse of water during the application of the straw, which indicated that caution should be used in respect of excessive wetting in any further rethatching.

SPECIFICATION FOR RETHATCHING AND REPAIR WITH LONG STRAW, BERG COTTAGE

Note: This specification was written by the thatcher who also conducted the survey of the roof. A simple specification is appropriate if the contractor is known to the client as an experienced thatcher and is tried and tested.

Specification

1. Remove all old netting and cart away.

2. All excessive thickness, wet and badly decayed thatch, to be removed and carted away.

   The eaves and gables <u>are not</u> to be completely stripped as a matter of course except for areas that are wet and badly decayed (see note 3). Give details of the adjustment to the old existing coat work. Any base coat work being over thatched, must be level and well compacted for the correct overlaying of the new material.

3. Where gables lack sufficient overhang to protect the wall or end rafter, they should be replaced. Provide details (see note 5).

4. The whole roof should be over thatched in long straw to an overall thickness of not less than 250 mm, with a minimum cover of 130 mm over the fixings. Every course is to be fixed with hazel spars and all dimensions should be consistent across any given roof area.

5. The thatch is to have a minimum overhang of 50 mm to all gables. All eaves and gables should be over fixed with hazel spars and liggers, cut square.

6. The finished roof should be well fixed and of even density, texture and pitch.

7. The ridge is to be of long straw and of a flush wrap-over type with pinnacles to all gable ends. The wrap-over is to be a minimum thickness of not less than 110 mm. The ridge should be well fixed with spars and liggers and follow the pitch of the underlying roof.

8. The completed roof is to be covered in 19 mm, 20 gauge galvanised wire netting. The netting is to be twist jointed at the ridge and all other edges. The netting at the eaves and gables is to be tucked over the tilting filet to form a neat finish. All eaves, gables and exposed rafters are to be made vermin proof.

9. Provide lead flashing to gable walls at both ends of the property and stepped flashing to the chimney stack. All flashings are to be of Code 4 lead and the appropriate precautions are to be taken if joints are to be burnt (lead work is subject to Listed Building Consent).

Materials

A suitable variety of wheat giving adequate length and strength should be used. Maris Huntsman and Maris Widgeon are common but other traditional varieties should not be excluded.

The straw should be cut with a binder and stooked to ripen, before being passed through a thrashing drum. The straw should not be too crushed or broken and should be in such a condition so as to allow for preparation into yealms. Excessive use of water should be avoided in the preparation of the straw prior to thatching. The straw should be fibrous and not brittle. A hollow-stemmed straw of a healthy bright appearance with no discoloration should be sought.

Details of any new fitting or replacement of roof timber should be given.

Special notes

The appropriate ventilation to the roof void must be maintained in accordance with the recommendations set out by the Building Research Establishment but there is a requirement within the project to reduce excessive draughts.

Detail the use of scaffolding or any extra plant other than ladders.

All work must comply with Health and Safety Regulations and appropriate certificate and licences must be obtained before commencement of any works.

Any change of materials, finish or detailing to the thatch and the underlying roof structure of a listed building will require Listed Building Consent.

Specification author: Keith Quantrill

COMMENTS ON THE RESULT

It should be noted that a much wider variety of traditional straw types is now available since this work was undertaken.

The consultant employed established a consistent basis for tendering and pricing the rethatching and necessary repairs to the building, which allowed for competitive prices to be received from a number of companies.

The key approaches which enabled the success of the project were:

- conducting a detailed thatch condition survey and assessment
- conducting historic research into the building and thatch archaeology to enable correct materials, detailing and aesthetic to be achieved. This enabled the Listed Building Consent to be received for change of detailing to the ridges and verge
- assistance from the consultant to assess and select an appropriately experienced and competent thatching contractor to undertake the works
- tendering process undertaken based on a specification and site visits by the consultant, which allowed contractors to price on similar basis of works and for there to be a clear understanding from the outset of the complete package of works and intention of the client
- selection of the contractor was undertaken based on competitive pricing and assessment by the consultant of their competence, programme and ability to fulfil the works as intended

- management of the works from commencement to completion by an independent consultant enabled material selection to be approved for quality and monitoring of all aspects of the works to safeguard the client's interests. Material assessment was approved following visits to the supplier and source to agree the condition of the materials.

As the works were part of a larger piece of conservation work, a close working relationship between the thatcher and conservation builder was important in managing the details around the perimeter of the building and linked junctions at abutments.

Selworthy Cottage, Holnicote Estate, Somerset

The Holnicote Estate comprises 5042 ha of moorland, farmland and woodland. It borders the coast along five miles of the Bristol Channel and rises to 519 m at Dunkery Beacon, the highest point on Exmoor. The whole estate is part of the North Exmoor Site of Special Scientific Interest (SSSI). It includes the villages of Allerford, Bossington, Luccombe and Selworthy and several hamlets. There are 167 cottages, 14 farmhouses and associated farm buildings. The major part of the current estate was passed to the National Trust in 1944.

Thatching work at Holnicote has set the standards to be adopted within the Trust for thatching works. Some 67 of the buildings are thatched, including seven barns. In 1999, a survey of the thatched roofs was carried out which identified that a large number were in very poor condition (an example is shown in Figure 5.6), resulting in tarpaulins being used to prevent water ingress into the buildings. This provided the impetus to set up a dedicated thatching project on the estate to gain a better understanding of thatching techniques and materials used. Its primary objective was the repair and rethatching of the leaking roofs, but it also enabled a greater appreciation to be developed of historic thatching traditions in West Somerset, the methods and materials used.

A number of roofs identified as being of historic interest were examined archaeologically. With this knowledge, the Trust was able to re-establish the cultivation of specific kinds of wheat on the estate and to train an apprentice thatcher in local methods of thatching in combed wheat reed.

COMBED WHEAT REED

Combed wheat reed (CWR) is traditionally used in the south-west of England as a method of thatching (as opposed to using long straw or water reed). This method of thatching was historically employed at Holnicote and is still in use today. Although laid with the butts of the stalks exposed in a similar manner to water reed, a roof thatched in CWR has a close-cropped appearance, with the eaves and gables cut to shape, which distinguishes it from other types of thatch.

The method of preparing CWR for thatching is a labour-intensive process. Once the wheat is ready for harvesting, it is cut with a binder, stooked and left to dry in a rick until ready to be combed using a threshing machine or comber. The comber removes the grain and leaves from the wheat without the need for the stalk to go through the drum, thus leaving it undamaged. The straw is delivered from the comber with the butts laid in one direction ready for tying into bundles. At this stage the wheat is known as CWR and, once the bundles are trimmed, is ready for laying onto roofs.

Roofs thatched with CWR are not usually completely stripped prior to rethatching, thus leaving layers of older thatch in place. This technique of leaving historic materials *in situ* has provided the basis on which to carry out important archaeological recording of the roof covering and structure.

Figure 5.6 Threshing process for combed wheat reed. © National Trust.

Figure 5.7 A trimmed bundle of combed wheat reed. © National Trust.

CONTEXT

Selworthy Cottage is a Grade II-listed 19th-century cottage. The original 1820s structure was a two-storey 'L' shaped dwelling. A rear (north) extension was added in the late 1870s, part of which is tiled. The axis of the main roof runs east–west, with two asymmetrical bays extending south towards the road at the east end; a smaller western bay forming a two-storey porch encroaches onto the larger eastern bay to create a complex system of valleys at roof junctions.

A detailed thatch excavation was commissioned in March 2000 to assess the site and building, the condition of the thatch, the ecology of the covering, the roof structure and the base coat as viewed from the underside within the roof void. A record of the excavation was compiled, together with an analysis of the subsequent findings, a list of samples taken and a photographic record.

The step trench was excavated on the east face of the bay and revealed four distinct layers of thatch. Analysis determined the following.

- Evidence from within the roof space suggests that the roof had undergone two major phases of reconstruction.

Figure 5.8 Selworthy Cottage before work began. Note the condition of the ridge. © Historic Thatch Management Ltd.

- Some of the original thatch and fixings remained on the underlying ceiling in the roof void.
- The original base coat was stripped from the entire roof, probably when the rear (north) extension was constructed.
- The surviving chronology indicated that there have been at least five phases of rethatching since the base coat was replaced.
- This build-up of thatch only accounted for approximately 70 years of thatch life. It is very likely that at least one, and probably two, other spar coats were applied (and subsequently stripped) at some point.

CONSERVATION PRINCIPLES APPLIED

Today's techniques and materials were similar to those used in past rethatching.

- Weathered surface layers were stripped to a sound base, leaving patches of moss and lichens *in situ*.
- Eaves lines were stripped and bottles retied to the lower most batten to provide a secure foundation.
- New CWR was sparred into place with split hazel and willow spars and sways (horizontal straw bonds were used at the last rethatching).
- The surface was dressed into position using a dreft (leggett).
- Hips and eaves were trimmed to shape.

Finally, a ridge roll was added to raise the ridge to the height of the new main coat, prior to the reinstatement of a butts-up ridge.

SPECIFICATION FOR RETHATCHING WITH COMBED WHEAT REED, SELWORTHY COTTAGE

1. General items

1.1 All personnel working on site must be aware of, and abide by, the guidelines set out in the 'General Requirements for Building Works' published by the National Trust.

1.2 Historic materials and fixings will be retained and consolidated whenever possible.

Where stripping of historic materials is inevitable, examples will be bagged, labelled and stored for later analysis.

No work will be undertaken on roofs with smoke-blackened base coats without prior approval of the National Trust regional building surveyor.

1.3 A photographic record will be kept of all rethatching work, including before and after photographs of all elevations, and detailed photographs of historic materials exposed and the range of techniques used in the rethatching.

2. Materials

2.1 Timber: All replacement timber will be supplied from the Holnicote Estate by National Trust staff unless an outside contractor has been specifically authorised to supply material by the assistant building surveyor.

2.2 Straw: All rethatching is to be done using combed wheat reed (CWR), preferably derived from suitable winter wheat varieties grown in organic conditions and approved by the Estate thatcher.

Preference will be given to reed grown locally, preferably on the Holnicote Estate or in Somerset or adjacent Devon.

All straw will be approved on delivery by the Estate thatcher and a National Trust representative, and a sample of each delivery will be labelled and stored for future analysis.

All growers supplying straw for use on the Estate must submit a completed 'Thatching Straw Production Record' sheet prior to receiving payment.

2.3 Spars: All spars and sways are to be made from hazel or willow and, wherever possible, are to be cut from Holnicote Estate forest by National Trust staff.

Both split or round wood spars and sways will be used as approved for the purpose by the Estate thatcher and a National Trust representative.

2.4 Cord: Only hempen or flaxen cord will be used for tying in wads (bottles) or any other purpose.

3. Works practice

3.1 Eaves: Wads are to be tied individually on eaves and gables using cord, with every other wad being sparred to its neighbour.

3.2 Fixings: All coat work is to be fixed either with horizontal sways or straw bonds, tied or sparred as required.

No iron crooks or wire fixings are to be used except with permission of the Estate thatcher and the National Trust regional building surveyor.

3.3 Ridge: All ridges are to be finished flush, in the butts-up style, secured externally with two horizontal liggers and spars.

All finials to be West Somerset 'points' with single cross spars between the liggers.

All other forms of external detailing to be agreed with the National Trust regional building surveyor.

3.4 Trimming: After the final dressing, the thatch surface is to be hand trimmed using a shearing hook.

Eaves are to be cut with both hook and shears.

4. Netting

4.1 Netting is to be used only with the prior approval of the National Trust regional building surveyor.

5. Mortar and flashing

5.1 Lime mixes will be used for all mortar work and will be supplied only by National Trust staff.

Specification author: John Letts, Historic Thatch Management Ltd

COMMENTS ON THE RESULT

For future repairs, the local surveyor would not change the approach that was taken in rethatching buildings on the estate. However, many of the buildings are listed and so additional requirements of the local conservation officer had to be considered which resulted in some sacrifices being made regarding traditional materials for perceived improvements in durability; for example, instead of using estate-grown and therefore sustainable chestnut for timber roof repairs, oak had to be bought in from outside the area and earth cob used for repairing walls had to be made up with the addition of lime, which is not normally added in this locality.

Appendix 1
Conservation management plans and conservation statements

A **conservation management plan** (CMP) is a systematic evaluation of the significance of a place in order to establish where that significance is vulnerable to threat. It uses the information to develop policies and subsequent actions to guide sustainable management of the significance. It addresses the following questions.

- What is it?
- Why does it matter and to whom?
- What is happening to it?
- What are we going to do about it?

Preparation of a CMP is participatory, requiring consultation with wider communities of interest than just specialists to ensure that the full range of values of a place is understood and to demonstrate best practice in managing change. Values may be historic, scientific and social, and relate to both tangible (i.e. the fabric of a place) and intangible heritage (i.e. other values that people ascribe to a place, such as memory, tranquillity or stimulation). See Appendix 1, Annex 1 below for a CMP checklist.

The CMP may be needed:

- to identify and understand significance
- to identify all the interests involved
- to assess the impact of change
- to help integrate and inform different policies
- to inform surveys and research.

A **conservation statement** is a useful shorter, summary version of a CMP that addresses:

- the significance of the place based on current levels of understanding
- current proposals for the place
- any other current or anticipated issues that might affect the place. A SWOT analysis (Strengths, Weaknesses, Opportunities, Threats) can be valuable for this part of the process
- how these issues will be dealt with

If there is uncertainty about whether a full CMP is needed, an initial conservation statement is recommended as this may be all that is required, or it may recommend further research or survey or a full CMP. It follows the same structure as a CMP.

A conservation statement might also address needs and help to guide decisions on:

- whether to address development proposals in-house through the project team
- prioritisation of future conservation programmes, including repair, refurbishment or reuse
- requirements for future research or survey
- strategies for access and interpretation
- informing and supporting the CMP
- informing stakeholders and any future consultation
- identification of funding needs to carry a project forward

Either a CMP or a conservation statement may be applied to a whole property or to individual elements, for example a house, garden, park or other landscape component. They can also be used for individual objects, such as a water feature, carriage or a collection, as well as cultural or natural sites.

STAGES OF A CONSERVATION MANAGEMENT PLAN

1	• Agree the aim of the plan and how it will be used • Involve people – identify who has an interest
2	• Understanding and analysis • Identify the different types of heritage involved • Include context
3	• Why is the place important? • Assessment of significance
4	• Explore issues and opportunities • SWOT analysis
5	• What do you want to achieve? • How will you manage the heritage? • Set policy objectives for managing the asset
6	• What do you need to do? • Management actions • 'As much as necessary and as little as possible'
7	• Use, monitor and review the plan • Use the plan – do it! • Review it! • Keep the plan up to date

FRAMEWORK AND CONTENTS

The CMP or conservation statement should follow the template laid out opposite.

Content list	Explanation

1. Introduction and summary
 	
  1.1 The aims of the plan
 	
  1.2 The need for the plan
 	
  1.3 Scope and status of the plan
 	
NB. *Where a plan is being revised, include an evaluation of the previous plan here.*
	The plan should start with an executive summary explaining why the CMP or conservation statement is needed and how it will be used; for example, to establish properly balanced priorities for the management of a complex property or part of the property.

2. Understanding the place or asset (description and illustrations of the property or place)
 	
  2.1 Location and description
 	
  2.2 Setting and context
 	
  2.3 Historical development
 	
  2.4 Designations
 	
  2.5 Gazetteer (inventory)
	It is important to identify what information is already available and whether new surveys are required to fill gaps in knowledge. This section may include all or some of the following, and should be supported by plans, including phasing plans and other illustrations.
	
	• **Ownership, acquisition** and **location**.
	• A **description of the whole site or property**, listing its main components, whether landscapes, buildings, ecology and biodiversity, archaeology etc, and their inter-relationships, both chronological and spatial. It is important to keep the core plan as short as possible. Detailed information should be held in the 'gazetteer', which can be included as a separate appendix to the main plan.
	• Where appropriate, **landscape character areas** should be identified and described. These can be used to further subdivide the plan if necessary. They should also be linked to the gazetteer, so that components for each character area are grouped together.
	• A **description of the geomorphology, soils, water resources and other natural processes**.
	• A **history of the property or place**, including a description of how it has changed through time and what survives from each phase.
	• A **description of its function**, how it worked and the people and processes involved.
	• An **assessment of the wider context** for the place and its physical setting, whether in ownership or beyond the boundary.
	• An **indication of where information is inadequate** and where further information (e.g. survey and research) may be required for the development of current proposals or for enhanced understanding to guide ongoing management.
	• A referenced list of **previous research, recording and analysis**.
	
	Detailed data on the various categories of information should be held in a separate **gazetteer** (see Appendix 1, Annex 2), which enables key information about each component part of the property or place to be easily accessed. The components may be individual archaeological sites, specific habitats, buildings or parts of buildings, landscape components, including parks and gardens, or objects or group of objects.

Continued

Content list	Explanation
	The gazetteer will include a summary description, any designations and relevant geographical data, and also include a short statement of why that particular component is significant in the context of the whole property. It should also include key issues and management recommendations affecting that particular component.
	Detailed surveys, research statements and gazetteers can be included as appendices to the main plan, within separate volumes if necessary. For example, archaeological, ecological or landscape surveys should form appendices, in order to keep the core document as short and accessible as possible. For a complex property, individual gazetteers relating to specific surveys can also be delivered as separate volumes or appendices.
3. Assessment of significance	This section follows on from the previous descriptive account of the place and will include an overall statement of the importance and significance of both the physical character of the place and also other less tangible values and attributes.
	Identify all the statutory, national or local designations that apply to a place, including references to any relevant local policies in the local plan.
	Explain why the place is important, including the significance of its component parts and its wider landscape setting and other contexts. Ensure that all potential areas of significance have been considered, including, *inter alia:*
	• Cultural significance: features and processes, such as buildings and artefacts, archaeological, historical, architectural and art historical evidence, collections, craftsmanship, scientific and technological interest, horticulture, cultivated plants, domesticated animals, design, present and past usage. It is also important to include the cultural context of the place. • Associations with famous people, events, art, literature and music, myths and legends. • Environmental significance: natural features and processes such as plants and animals, ecology, geology, geomorphology, landforms, soils, hydrology, air quality, climate change, decay and ageing. • Social significance: access, including the value of the place for all users and visitors, opportunities for leisure and learning. • Economic significance: socio-economic activities such as farming, forestry, tourism, recreational opportunity, transport, local business, industry, employment, and development. • Other values, relating to 'spirit of place': it is also important to consider special characteristics such as individuality, distinctiveness, typicality or representativeness, group value, rarity or uniqueness, and condition and quality of survival.

Content list	Explanation
	• Also include the less tangible values experienced by people who know or have an involvement with the place or asset, including, for example, aesthetic qualities, such as setting, design, style, scale, character, naturalness, wildness, and awareness of by-gone times or 'time depth', i.e. its antiquity, and other emotions and spiritual qualities such as awe, beauty, delight, evocation of memories, excitement, peace, sense of involvement and participation, surprise, tranquillity, etc.
4. Identification of key management and conservation issues and opportunities	A key part of the conservation planning process is to identify where the new opportunities exist for enhancing the significance of a place or where that significance is vulnerable to threat and impacts need to be analysed. A SWOT analysis can help with this process.
	This section will include an analysis of future needs, options, vulnerability and constraints. It needs to consider *inter alia* some or all of the following points and any others of relevance that have been identified.

	• Internal management and staffing structures and processes and how they are applied.
	• Options for future management, change or development.
	• Any external management constraints, e.g. tenancies, third party management.
	• Requirements for museum accreditation.
	• Loan agreements for collections.
	• Resources.
	• New use or an adaptive reuse for a place.
	• Services required.
	• An assessment of the vulnerability of the place including such factors as:
	  ❑ condition
	  ❑ current uses, including access, whether open to the public or not, impact of visitors and health and safety requirements
	  ❑ natural decay
	  ❑ pressures for internal development and change – what impact will that have
	  ❑ external pressures, for example, planning or development pressures adjacent to or in the setting or vicinity of the place
	  ❑ gaps in existing knowledge
	  ❑ identification of management risks to the place.

	The use of Heritage Impact Assessments (recommended by the Heritage Lottery Fund) may be required to:
	• identify the adverse or beneficial impacts each option will have on the significance of the place
	• identify what steps will be required to reduce or mitigate negative impact, or to assess whether the degree of impact causes such adverse loss of significance that a development should not go forward.

Continued

Content list	Explanation

5. Development of management policies and objectives
 	
  5.1 Current management context
 	
  5.2 Relevant organisational policies and principles
 	
  5.3 Local community
 	
  5.4 Planning framework
 	
  5.5 Management and conservation objectives for the site
	This section brings together relevant conservation policies, principles and guidance explaining how we will ensure that the significance of the place or asset is addressed in both ongoing management and in the context of any new development.
	
	It sets out the wider national and local statutory and legislative frameworks, policies, designations and regulations that need to be considered and may provide constraints on any development. It also develops detailed management policies for the conservation management of the place to address issues (identified above), mitigate the effects of adverse change and to achieve sustainable long-term management of the significant heritage assets.
	
	NB. Management policies will identify a course of action required to secure the conservation of the place or its components and will require certain actions to be undertaken. Policies can be set out as a strategic framework, i.e. relating to the long-term management of the place, and as short-term management objectives.
	
	More detailed short-term conservation objectives will flow from the policies described above. These objectives will relate to need or ability to achieve the requirements identified in the policies.

6. Management actions ('as much as is necessary and as little as possible')	These may range from day-to-day management of components, such as field boundaries or archaeological sites, to long-term remedial conservation programmes for collections. Resources need to be identified and activities prioritised and timetabled. It is important that all management actions flow from the issues, the relevant policies and conservation objectives – see table below.

Issues, risks	What is affected? Impact	Significance	Conservation plan policies	Management actions
Summary of the issue	Using the gazetteer, list parts of the property which might be affected	How will this affect the significance of the place?	List relevant policy(ies)	Summarise or list relevant management actions which have been identified

USE, MONITOR AND REVIEW THE PLAN

Keep the plan up to date – a five-year review period is sensible. The plan is a live document and needs to be reviewed periodically as new information becomes available and new values emerge. The review needs to include evaluations of the previous plan, including any new research or information that may affect the significance, and its policies, objectives and actions should be reviewed.

Regular processes of monitoring the condition of a place, for example, a quinquennial inspection of a building or tree surveys in a historic park, need to be fed back into the plan and fresh management actions need to be identified as necessary.

Appendix 1, Annex 1 Conservation plan checklist

Identify the type of conservation plan required		NEW UPDATE		
Do you have the required resources available?		Yes		No
Where will the resources come from and how much is available?				£
Task	By whom	By when	Completed	Comments
Task 1: Prepare the brief			Yes / No	
Task 2: Set out the timetable			Yes / No	
Task 3: Identify the author/s and other interested parties			Yes / No	
Task 4: Locations of additional information			Yes / No	
Task 5: Identify property areas			Yes / No	
Task 6: Evaluation of the site together with its significance			Yes / No	
Task 7: Analyse the features/ collection			Yes / No	
Task 8: Produce a priority list			Yes / No	
Task 9: Overview conservation plan			Yes / No	
Task 10: General conservation proposals for each defined area			Yes / No	
Task 11: Complete property policies			Yes / No	
Task 12: Programme of work including resources			Yes / No	
Task 13: Identify resources needed			Yes / No	
Task 14: Distribute for comments			Yes / No	
Task 15: Final draft			Yes / No	
Task 16: Set review time scale			Yes / No	

Please note some of these tasks might not be relevant to all conservation plans.

Appendix 1, Annex 2 Example of a gazetteer

Information that you might usefully include within a gazetteer.

- Area and reference number
- OS grid reference
- GIS reference
- Designations/statutory protection
- Overall significance of the component part or feature
- Justification of significance – a summary statement
- Date of inspection by specialist gazetteer compiler
- Current use, e.g. arable farmland, derelict building
- Area in hectares
- Significant elements:
  - reference no
  - type, e.g. Roman villa or hedgerow
  - brief description
  - significance
- Issues, e.g. vulnerable to ploughing, avoid use of insecticides
- References, e.g. to text on the main conservation plan or to published sources
- Reference to relevant policy objectives, which might relate to the component (it may not be necessary to include this in the gazetteer)
- Management actions necessary for the individual component

Appendix 1, Annex 3 Criteria for determining significance

DEFINITION AND ASSESSMENT OF SIGNIFICANCE

Buildings

Government guidance notes on the management of the listed buildings and conservation areas provide the following criteria for determining significance.

- Architectural interest – design, decoration, craftsmanship, building type, technique.
- Historic interest – illustrate important aspects of the nation's social, economic, cultural or military history.
- Historical associations – with important people or events.
- Group value – where buildings comprise an important architectural or historic unity, or constitute a fine example of planning.

Archaeology

Government guidance notes on the management of archaeology provide the following criteria for determining the significance of ancient monuments and archaeological sites.

- Period – monuments that characterise a category or period are significant.
- Documentation – significance enhanced if there is supporting evidence.
- Group value.
- Survival/ condition.
- Fragility/vulnerability – more vulnerable monuments benefit from greater protection.
- Diversity – a combination of features.
- Potential – when features are anticipated but the extent is not yet known.

Collections and archives

There are various criteria which need to be considered when assessing the significance of an object or a collection, including their museum status where appropriate.

Objects in museum collections will be subject to the terms of Museum Accreditation. The Museums, Libraries and Archives Council has published criteria for the definition of important objects defined as pre-eminent under its Acceptance in Lieu Procedure. These are that the object:

- has an especially close association with our history and national life
- is of especial artistic or art historical interest
- is of especial importance for the study of some particular branch of art, learning or history
- has an especially close association with a particular historic setting.

In addition, the following criteria should also be considered.

- Rarity.
- Quality of the design and skill of craftsmanship.
- Historic interest of use or associations with important individuals or with the place.
- Unity with other aspects of the property such as buildings or landscapes.
- Documentation – the extent to which the provenance can be proved.
- Condition.

Landscape

Historic England defines the following criteria for assessing the significance in its Register of Parks and Gardens of Special Historic Interest.

- Age.
- Influential/famous sites.
- Important designer.
- Good examples of its type.
- Associations with nationally important people or events – requires a direct link, which is reflected in the layout of the site.
- Group value – with buildings or other land.
- Interest of phases of development.
- Documentation.
- Condition.

Biodiversity

In its guidelines for the selection of Sites of Special Scientific Interest (SSSI), the Joint Nature Conservation Committee includes the following criteria for assessing ecological significance.

- Size – the physical extent of the habitat.
- Naturalness – extent to which habitats are unmodified and free from artificiality, and the species it contains are typical of the habitat.
- Rarity and fragility – number of species which are nationally rare, prone to extinction, or are listed in the Red Data Book.
- Diversity of species – the number of different species and the extent of the habitat mosaics where species depend on more than one habitat.
- Population size of species.
- Special Areas for Conservation (SACs) – these are all SSSIs that meet rigid criteria for European importance, and include rare biological communities.
- Geo-diversity. It is also recognised at national level through the SSSI system, against very rigid criteria.

… and much more so:

- Documented history.
- Biological potential.

Appendix 2
Generic brief for a historic buildings survey

This generic brief is devised not just to provide a record of historic fabric and structure, but to include an analysis and interpretation of the building's origins and development and assessment of its significance. It should also include, as far as possible, an understanding of how the building was used and the processes involved in its development and use.

In view of the possible changes that may be proposed in the future, recommendations should be made for practical and effective management that will not compromise the building's special features or overall conservation value.

For each building surveyed, the completed report will provide the most recent digital record, capable of being added to or enhanced as need may arise. It will be an essential tool for staff in planning processes and is also a good stimulus to responsible stewardship by tenants and other occupants.

Refer to Historic England's *Understanding Historic Buildings: a guide to good recording practice*. This gives full details of the most generally applicable building recording techniques and includes a discussion of appropriate levels of survey and a set of drawing conventions: www.historicengland.org.uk/images-books/publications/understanding-historic-buildings/.

The survey brief

1. INTRODUCTION

The brief should be drawn up by an appropriately qualified person and tailored to the character of the building to be surveyed and to any relevant circumstances (e.g. proposed alterations) prevailing at the time. It should also reflect the particular level of survey required; fully researched and detailed measured and analytical surveys may not be appropriate in every case and in some instances summary descriptions and digital photographs will suffice. The introduction to the brief should include the following.

- The historical and archaeological background of the building.
- Location and description of the property.
- The current tenurial background.
- Particular reasons why a survey is required, e.g. vacant building with proposals for new lease or for adaptive reuse.
- The number and location of the building to be covered by the survey.
- Summary of any previous research and recording.
- The scope and level of survey required.
- Timetable for the recording project.
- Invitation to the contractor to prepare a costed project design or written scheme of investigation.

2. PURPOSE OF SURVEY

2.1 The purpose of the survey is to assess the origin, construction and development of the building with the particular intention of establishing its significance both as a discrete individual structure and with regard to its relationships as part of a historic group or landscape.

2.2 The resulting report will provide a metrically accurate building survey with a comprehensive visual record and a report with a focus on interpretation, presentation and understanding. This will inform the nature of any future repairs and any conservation measures that might be needed to safeguard the historic fabric.

2.3 It will also provide a baseline digital record against which any future changes to the building or alteration to the management of the property may be measured and recorded.

3. BACKGROUND RESEARCH

3.1 *Listed Buildings and Scheduled Ancient Monuments:* a copy of the list entry and other related statutory designations should be obtained and included as an appendix to the report. It would also be helpful to note whether a building is within a Conservation Area and any other relevant designations, such as AONB or National Park.

3.2 *Sites and Monuments Register or Historic Environment Record:* any information held on the county or unitary authority register relating to the building and its site, or immediate vicinity, should be obtained. This will help to provide an understanding of the wider landscape context of the building.

3.3 *Documentary research*: a copy of all maps and plans held in the Record Office, local history library or other relevant record office, showing the buildings should be copied, with their date, scale and reference recorded. Other historic illustrations, paintings, prints or early photographs should also be collected.

3.4 *Further documentary research*: written material is almost always valuable. Documentary research is essential for interpreting the fabric and to throw light on the building's origins, past uses and its social and historical context. All sources must be fully referenced.

4. FIELD METHODOLOGY

It is important to note that active interpretation and analysis during field work are important, rather than waiting until after the survey is complete. Very often, the opportunity to question a possible interpretation only arises when making a field observation.

4.1 Drawings
Refer to the drawing conventions in *Understanding Historic Buildings.*

An accurate, drawn, scaled record should be produced and held digitally (for example, on CAD) to give plans of all floors, plus cross-section(s) at suitable point(s) to show vertical relationships and roof construction, and where necessary for understanding, a cross-section or long section, elevational and detail drawings. The survey will examine and check the validity of any existing analysis and add further detail as needed. Plans should include the following details.

- All openings.
- All blocked openings.
- Structural relationship where clear.
- Permanent fixtures (e.g. fireplaces, fitted cupboards, etc.).
- The position of structural timbers, including overhead beams.
- Any other significant features.

4.2 Written descriptions
These should include the following.
   (i) Exterior

- Roof, shape, covering materials, style
- Walls, materials, any blockings, lintels, recesses, abutments/relationships to other walls/ structures, features (e.g. bread oven bulge)

- Chimneys – materials shape, pots, etc.
- Anything else of relevance

  (ii) Interior – room-by-room record referenced to numbered or annotated plan including:

- ceilings
- walls
- floors
- fireplace
- doors, windows and other fixtures
- roof structure
- machinery, or evidence for its former existence.

4.3 Photographic record

A fully indexed photographic record is required to support drawings and descriptions. This will include digital photography and, where necessary, 35 mm monochrome photography. Resolution of digital photography will need to be specified and should be no less than 300 dpi. Publication quality photographs will be 600 dpi.

A record should be made of the location and direction of photographs taken.

Coverage will include:

- photographs showing relationships between elements of a building, including general room interiors
- where appropriate, elevations, internal or external, particularly if the building has a complex decorative scheme which needs careful recording
- individual features of interest, e.g. fireplaces, stairways, decorated ceilings and fittings, including distinctive door and window furniture and other architectural decorative features.

4.4 Specialist survey

The following may be recommended.

Dendro-dating
- Use of dendrochronology for potential dating of structural timbers and joinery
- Retention (and marking) of offcut samples for potential future dating
- Retention and archiving of samples used by specialists

In the case of potential dendro-sampling and analysis, it is essential that a curator, archaeologist and/or conservator confers on site with the consultant archaeologist and dendrochronologist. This is to establish whether those areas of timber that are physically capable of potential results are in locations that have been approved by staff for sampling. It may also be necessary to gain consent from Historic England (scheduled and listed sites). Areas proposed for sampling will not normally be in main show-rooms or on the visitor route where collateral damage to the historic fabric might be incurred in the sampling process. Wooden plugs used to 'make good' timbers that have been selected for sampling must be on a like-for-like basis (e.g. native oak sample to be replaced with native oak plug). They should ideally be of similar type and must be from a reputable renewable and sustainable source.

Paint research
- Use of paint research to establish former decorative schemes
- Geological and/or chemical analysis of stonework or mortar
- Paleo-environmental analysis of, for example, thatching materials

5. REPORT

An example of a pro forma report structure is shown in Appendix 2, Annex 1 below. The report will include all or some of the following.

- An abstract or summary.
- An introduction, covering the location and setting of the building, and the circumstances, scope and methodology of the survey; also acknowledgements.
- The historic background to the site and the building(s), including historic maps and illustrations. This should include its historical relationship to other buildings in the vicinity, their uses and their related processes and the wider landscape context.
- Description of the building(s) as existing and inventory of details. This should include the interior and exterior of the building and a description of its current relationship with others in the complex.
- Interpretation and significance of the building. This will include an analysis of the architectural structure, historic phasing, changing function and dating.
- An analysis of current condition and risks.
- Key conservation and management objectives (where appropriate). These should be determined in consultation with professional specialists. Discussion will involve the curator more actively in assessing the management implications of any fresh understanding of the building that the new survey has produced.
- Where required, the report should include those features that have been identified as crucial to the historic integrity of the building and that must not be lost, and any that might be less significant. This should be practical guidance that is capable of application (rather like an impact assessment).
- Sources and bibliography.
- Figures, including copies of CAD drawings.
- Plates and an index to the photographic archive .

6. ARCHIVE DEPOSITION

6.1 The survey archive, including drawings, photographs, and research and field notes will be properly catalogued and deposited with the owner.

6.2 Report procedure and distribution

- Copies of the final copy of the report will be produced in a bound, double-sided A4 format. Digital copies will also be submitted in Word or RTF format on CD.
- Copies of the report will be provided to the client.

7. SURVEY PROGRAMME AND PROJECT REVIEW ARRANGEMENTS

7.1 The dates for submission of reports will be agreed between the owner and the contractor, but the normal assumption is that a draft report will be submitted within one month of the final site visit for each building or complex of buildings.

7.2 A final report, complete with project archive, will normally be submitted within a month of approval of the draft.

7.3 Invoices will be submitted as agreed on a phased basis, normally on completion of the survey for each building or group of buildings.

7.4 If VAT is payable, this too should be included in the bid.

8. ARCHAEOLOGICAL PROJECT RECORDING FORM

A completed form will be submitted by the contractor with the final copies of reports (see Appendix 2, Annex 1).

9. COPYRIGHT

9.1 The owner will retain full copyright over information, reports and plans, and shall have absolute control over the use and/or dissemination of that information which may not be published in any form without the owner's consent. The owner will not unreasonably withhold such consent.

9.2 The contractor will be fully accredited wherever the material is used or reproduced.

10. PUBLICATION

The contractor should produce a short summary of the report, including the building's overall significance, history and significant features. This should be in a format that can be submitted to relevant national or local journals for inclusion in the annual notes section. Relevant journals could include *Medieval Archaeology*, *Post-Medieval Archaeology*, *Vernacular Architecture*, and appropriate county journals. Where appropriate, a full published report should be considered by the owner or their project staff.

11. HEALTH AND SAFETY

11.1 The contractor will take sole responsibility for observing all current legal requirements concerning their or their employees' health and safety. The contractor will supply the owner with a copy of their health and safety policy on submission of their quotation.

11.2 The contractor will also be required to provide a copy of a risk assessment for the survey.

12. INSURANCE

The contractor will be required to demonstrate that they hold public liability insurance to the value of not less than £2,000,000 and £5,000,000 professional indemnity insurance.

13. GENERAL TERMS

13.1 Contractors commissioned to undertake historic building recording and analysis should be able to demonstrate appropriate qualifications and experience as part of their tender document

13.2 All work will be undertaken to the Chartered Institute for Archaeologists' *Standards and Guidance for the Archaeological Investigation and Recording of Standing Buildings or Structures*; see www.archaeologists.net/codes/ifa

13.3 Appropriate procedures under the relevant legislation must be followed in the event of the discovery of any artefact covered by the provisions of the Treasure Act 1996. The owner/project manager must be informed and the find must be reported immediately to the local coroner.

13.4 In the event of the discovery of human remains, the owner/project manager must be informed immediately, and the local police and the coroner must be contacted.

13.5 The project will be undertaken by the contractor acting on an independent basis. Staff working on the project will not be deemed employees of the owner. Tenders should reflect this fact and more specifically, the contractor will take sole responsibility for the payment of tax, National Insurance contributions, etc.

14. CONTACTS

This should include the names, addresses, email addresses and fax and telephone numbers of relevant contacts.

Appendix 2, Annex 1 Pro forma report structure example

INTERPRETIVE HISTORIC BUILDING SURVEY

Site name

Commissioned by
[name of owner]

[photo of building]

Project no.

[year]

Prepared by

With contributions from

Name and address of commissioned archaeological contractor

HISTORIC BUILDINGS SURVEY

Summary Sheet 1

Property Name	Address	Building Name
Property/Building Reference	OS Grid Reference	Surveyor/Date of Survey
Category	Original Use	Current Use
Date(s) of Construction	Statutory Designation(s)	Sites & Monuments Record Reference
Walling Materials	Roofing Materials	Flooring Materials

Description:

Architectural/Historic Significance:	Landscape Significance:

Notes/Qualifications Regarding Survey:

Additional Information Sources for this Building (s):

Copies and CDs of this report held at:

Summary Sheet 2

Property Name	Address	
Property/Group Reference	OS Grid Reference	Surveyor/Date of Survey
Local Planning Authority		Local Authority (Building Regs, etc.)
Area Designations		
Description of Group (and any related buildings):		
History and Development		
Information Sources		
Maps		
Written Records		
Owner Records		

Contents

Preliminary pages

Abstract

List of figures

  Figure 1 Location plan

  Figure 2 Site plan

List of plates

  Plate 1

1. Introduction

To include name of site/building, general description of type of building and location plus whom the work was commissioned by and when. Ownership details of the site/building. State if building is listed, give listing and refer to appendix where list entry should be presented. Refer to location plan and property ownership boundaries if relevant.

2. Scope and methodology

Description of the extent of the work undertaken – that is, if the building is part of a group, were all the buildings examined or only one, what was examined? Any existing survey information utilised in the survey, e.g. previous plans digitised, and vernacular building survey data. Refer to site plan to inform text on scope.

Description of the method and level of survey undertaken – outline plan prepared/ detailed measured survey, etc., room-by-room description, photography – digital, black and white, colour transparency. Describe how and where the data are stored. (See Historic England's *Understanding Historic Buildings* for details of survey levels. Most detailed surveys of historic buildings will generally be at Level 3, but buildings of little historical significance will often only require Level 1 or 2 surveys.)

Description of documentary research, if undertaken. If not undertaken, state that and explain why. For documentary research carried out, describe the types of records examined and their location. Refer to section 8 of the report where the sources consulted are set out.

3. Location

Description of the location of the site. Give grid reference for site. Description of where it is, e.g. 0.5 miles south of village of xxx. Description of geology and geomorphology to establish landscape context.

4. Significance(s) of the building

State why the building is significant. Consider its history, design, construction, uses, its context including its relationships with wider processes, for example agricultural or industrial, and associations. Aesthetic significance may also be important. The importance and significance of individual features should also be referred to here.

5. Historic background

Give a general overview of the history of the area in which the building is located first, including, if it was part of an estate, giving any relevant data relating to that estate which may have a bearing on the understanding and interpretation of the site, and follow this with a more detailed description of the site of the building thereafter. Refer to figures. Use footnotes to give precise references to documentary sources.

6. Description of the building/s as existing

Give a general description of the building and surroundings.

Exterior – describe morphology and layout/orientation, building materials – use and distribution, elevations – fenestration/doors – give number, materials, shape, etc., rainwater goods – materials, location. Explain if elements are of interest and state why and whether they are dateable. Refer to plans and elevations.

Interior – describe interior of building – begin with an overview of structure, plan-form, circulation, and give an indication of characteristic or repetitive features (hierarchy of decoration). Then working from ground floor to roof, describe on a room-by-room basis, giving shape and size of room, structure where apparent, details of fixtures and fittings where relevant. Each room to be described consistently, e.g. floor, walls, ceiling, windows, doors, unless features throughout the building are ubiquitous, when an overall statement may be made at the start of the section rather than having to repeat it room by room.

7. **Interpretation of the historic phasing of the buildings**

On the basis of the description of the building as existing, set out the interpretation of the structure and the phasing. Give a general description at the outset and then subdivide the text into each phase, which should be numbered, and describe in detail and give reasons why the structure has been attributed to that phase.

8. **Key conservation and management objectives**

Set out general basic conservation principles for the protection of the significant features of the individual building. These may relate to construction, form and fabric, or to its use or potential for adaptive reuse. Include minimum intervention, like-for-like repairs, etc.

Specific conservation principles relating to the building should be included.

9. **Sources consulted**

List primary and secondary sources consulted, giving record office references where relevant.

Bibliography of specification-related conservation books and publications

Books

Allen, G, Allen, J, Elton, N et al. (2003) *Hydraulic Lime Mortar for Stone, Brick and Block Masonry.* Donhead Publishing Ltd: Shaftesbury.

Ashurst, J and Dimes F (1998) *Conservation of Building and Decorative Stone.* Butterworth-Heinemann Ltd: Oxford.

Ashurst, N (1994) *Cleaning Historic Buildings – Volume I – Substrates, Soils and Investigation.* Donhead Publishing Ltd: Shaftesbury.

Ashurst, N (1994) *Cleaning Historic Buildings – Volume II – Cleaning Materials and Processes.* Donhead Publishing Ltd: Shaftesbury.

Brereton, C (1991) *The Repair of Historic Buildings – Advice on Principles and Methods.* English Heritage: London.

Bridgwood, A (1951) *Carpentry and Joinery: Intermediate.* George Newes Ltd: London.

Brocklebank, I (2012) *Building Limes in Conservation.* Donhead Publishing Ltd: Shaftesbury.

Brown, R (1997) *Timber-Framed Buildings of England.* Robert Hale Ltd: London.

Brunskill, R (1997) *Brick Building in Britain.* Victor Gollancz Ltd: London.

Brunskill, R (1985) *Timber Building in Britain.* Victor Gollancz Ltd: London.

Brunskill, R (1977) *English Brickwork.* Ward Lock Ltd: London.

Brunskill, R (1976) *Recording the Buildings of the Farmstead.* Ancient Monuments Society: London.

Bryce, K and Weismann, A (2006) *Building with Cob – A Step by Step Guide.* Green Books Ltd: Totnes.

Charles, F (1997) *Conservation of Timber Buildings.* Donhead Publishing Ltd: Shaftesbury.

Clark, K (2001) *Informed Conservation – Understanding Historic Buildings and their Landscapes for Conservation.* English Heritage: London.

Eckel, E (2005) *Cements, Lime and Plasters.* Donhead Publishing Ltd: Shaftesbury.

Edlin, H (1991) *What Wood is That? A Manual of Wood Identification.* Stobart Davies Ltd: Hertford.

English Heritage (1997) *The English Heritage Directory of Building Limes.* Donhead Publishing Ltd: Shaftesbury.

English Heritage (2012) *Practical Building Conservation Series: Glass and Glazing.* Ashgate Publishing Ltd: Farnham.

English Heritage (2012) *Practical Building Conservation Series: Metals.* Ashgate Publishing Ltd: Farnham.

English Heritage (2012) *Practical Building Conservation Series: Mortars, Renders and Plasters.* Ashgate Publishing Ltd: Farnham.

English Heritage (2012) *Practical Building Conservation Series: Stone.* Ashgate Publishing Ltd: Farnham.

English Heritage (2012) *Practical Building Conservation Series: Timber.* Ashgate Publishing Ltd: Farnham.

English Heritage (2013) *Practical Building Conservation Series: Roofing.* Ashgate Publishing Ltd: Farnham.

English Heritage (2013) *Practical Building Conservation Series: Concrete.* Ashgate Publishing Ltd: Farnham.

English Heritage (2013) *Practical Building Conservation Series: Conservation Basics.* Ashgate Publishing Ltd: Farnham.

English Heritage (2014) *Practical Building Conservation Series: Building Environment.* Ashgate Publishing Ltd: Farnham.

Fearn, J (1995) *Thatch and Thatching.* Shire Publications Ltd: Princes Risborough.

Feilden, B (1995) *Conservation of Historic Buildings.* Butterworth-Heinemann Ltd: Oxford.

Harris, R (1993) *Repair of Timber-Framed Buildings.* Weald and Downland Open Air Museum: Sussex.

Henry, A (2006) *Stone Conservation – Principles and Practice.* Donhead Publishing Ltd: Shaftesbury.

Hewett, C (1980) *English Historic Carpentry.* Phillimore: London.

Hill, N, Holmes, S and Mather, D (1992) *Lime and Other Alternative Cements.* Intermediate Technology Publications Ltd: London.

Historic England (2006) *Understanding Historic Buildings: A guide to good recording practice.* Historic England: Swindon.

Historic England (2015) *Practical Building Conservation Series: Earth, Brick and Terracotta.* Ashgate Publishing Ltd: Farnham.

Historic Scotland (1994) *Conservation of Plasterwork.* Historic Scotland: Edinburgh.

Historic Scotland (1995) *Preparation and Use of Lime Mortars.* Historic Scotland: Edinburgh.

Historic Scotland (1996) *Thatch and Thatching Techniques – A Guide to Conserving Scottish Thatching Traditions.* Historic Scotland: Edinburgh.

Historic Scotland (1997) *Stone Cleaning – A Guide for Practitioners.* Historic Scotland: Edinburgh.

Historic Scotland (1999) *Stone Cleaning of Granite Buildings.* Historic Scotland: Edinburgh.

Holmes, S and Wingate, M (1997) *Building with Lime.* Intermediate Publications Ltd: London.

Howard, J (2009) *Bats in Traditional Buildings.* National Trust, English Heritage and Natural England: Swindon.

Insall, D (1972) *The Care of Old Buildings Today – A Practical Guide.* Architectural Press: London.

Ireson, A (1993) *Masonry – Conservation and Restoration.* Attic Books: Powys.

Jokilehto, J (1999) *A History of Architectural Conservation.* Butterworth-Heinemann: Oxford.

Jones, B (2003) *Building with Straw Bales – A Practical Guide for the UK and Ireland.* Green Books Ltd: Totnes.

Keeling, F (1956) *Carpentry and Joinery – Advanced Examples.* Cleaver-Hume Press Ltd: London.

Keeling, F (1959) *Constructional Carpentry.* Cleaver-Hume Press Ltd: London.

Kumar, A (2001) *Conservation of Building Stones.* Indian Council of Conservation Institutes: Lucknow.

Larsen, K and Marstein, N (2000) *Conservation of Historic Timber Structures – An Ecological Approach.* Butterworth Heinemann: Oxford.

Macdonald, S (1996) *Modern Matters – Principles and Practice in Conserving Recent Architecture.* Donhead Publishing Ltd: Shaftesbury.

Macey, F (2009) *Specifications in Detail.* Donhead Publishing Ltd: Shaftesbury.

McDowell, R (1980) *Recording Old Houses: A Guide.* Council for British Archaeology: London

Miller, W and Bankart, G (2009) *Plastering Plain and Decorative.* Donhead Publishing Ltd: Shaftesbury.

Moss, R (1988) *Lighting for Historic Buildings – A Guide to Selecting Reproductions*. John Wiley and Sons Inc: New York.

Nash, J (1991) *Thatchers and Thatching*. BT Batsford Ltd: London.

National Society of Master Thatchers (2008) *A Practical Guide to Thatch and Thatching in the Twenty-First Century*. National Society of Master Thatchers: Tetsworth.

Normandin, K and Slaton, N (2005) *Cleaning Techniques in Conservation Practice*. Donhead Publishing Ltd: Shaftesbury.

Oxley, R (2003) *Survey and Repair of Historic Buildings – A Sustainable Approach*. Donhead Publishing Ltd: Shaftesbury.

Pearson, G (1992) *Conservation of Clay and Chalk Buildings*. Donhead Publishing Ltd: Shaftesbury.

Plumridge, A and Meulenkamp, W (2000) *Brickwork*. Seven Dials: London.

Powys, A (1995) *Repair of Ancient Buildings*. Society for the Protection of Ancient Buildings: London.

Purchase, W (2009) *Practical Masonry*. Donhead Publishing Ltd: Shaftesbury.

Ridout, B (2000) *Timber Decay in Buildings – The Conservation Approach to Treatment*. Routledge: London.

Robson, P (1999) *Structural Repair of Traditional Buildings*. Donhead Publishing Ltd: Shaftesbury.

Rural Development Commission (1988) *The Thatcher's Craft*. Rural Development Commission: London.

Sawyer, J (2007) *Plastering*. Donhead Publishing Ltd: Shaftesbury.

Shacklock, V (2006) *Architectural Conservation – Issues and Developments*. Donhead Publishing Ltd: Shaftesbury.

Sharpe, G (1999) *Works to Historic Buildings – A Contractor's Manual*. Chartered Institute of Building: Berkshire.

Sloan, A and Gwynn, K (1993) *Traditional Paints and Finishes*. Collins and Brown Ltd: London.

Smith, B and Turkington, A (2004) *Stone Decay – Its Causes and Controls*. Donhead Publishing Ltd: Shaftesbury.

Smith, J and Yates, E (1991) *Dating of English Houses from External Evidence*. Richmond Publishing Company: Slough.

Sowden, A (1990) *The Maintenance of Brick and Stone Masonry Structures*. E and F Spon: London.

Smith, J and Yates, E (1968) *On the Dating of English Houses from External Evidence*. Volume 2 No 5. FSC Publications: London.

Strike, J (1994) *Architecture in Conservation – Managing Development at Historic Sites*. Routledge: London.

Swallow, P, Dallas, R, Jackson, S and Watt, D (2004) *Measurement and Recording of Historic Buildings*. Donhead Publishing Ltd: Shaftesbury.

Thatching Advisory Service (2003) *Patching and Re-thatching*. Thatching Advisory Service Ltd: Wokingham.

Thatching Advisory Service (2003) *Specifications on Thatch*. Thatching Advisory Service Ltd: Wokingham.

Tutton, M, Hirst, E and Pearce, J (2007) *Windows – History, Repair and Conservation*. Donhead Publishing Ltd: London.

Warland, E (1949) *Roofing*. English Universities Press: London.

Warland, E (2006) *Modern Practical Masonry*. Donhead Publishing Ltd: Shaftesbury.

Weaver, L (1928) *Tradition and Modernity in Plasterwork*. G Jackson and Sons Ltd: London.

Weaver, M (1993) *Conserving Buildings – A Guide to Techniques and Materials*. John Wiley and Sons Inc: New York.

Webster, R (1992) *Stone Cleaning and the Nature, Soiling and Decay Mechanisms of Stone.* Donhead Publishing Ltd: Shaftesbury.

West, R (1987) *Thatch – A Manual for Owners, Surveyors, Architects and Builders.* David and Charles: London.

Willis, C and Willis, J (2003) *Specification Writing for Architects and Surveyors.* Blackwell Science Ltd: Oxford.

Websites

Bats in Traditional Buildings: https://www.historicengland.org.uk/images-books/publications/bats-in-traditional-buildings/

Historic England energy efficiency guidance: https://www.historicengland.org.uk/advice/technical-advice/energy-efficiency-and-historic-buildings/

Historic England historic building repair and maintenance guidance: https://www.historicengland.org.uk/advice/technical-advice/buildings/maintenance-and-repair-of-older-buildings/

Historic Scotland energy efficiency technical papers: www.historic-scotland.gov.uk/index/heritage/technicalconservation/conservationpublications/technicalpapers.htm

Historic Scotland refurbishment case studies: www.historic-scotland.gov.uk/index/heritage/technical-conservation/conservationpublications/refurbcasestudies.htm

Society for the Protection of Ancient Buildings technical guidance: www.spab.org.uk/advice/

Sustainable Traditional Buildings Alliance: www.stbauk.org/

Index

Milton Keynes UK
Ingram Content Group UK Ltd.
UKHW022040141024
449569UK00014B/668